KT-366-046

FUNCTIONS OF A COMPLEX VARIABLE

I

BY

D. O. TALL

MEDICAL COMPUTING GROUP
LIBRARY

LONDON: Routledge & Kegan Paul Ltd

NEW YORK: Dover Publications Inc

First Published 1970
in Great Britain by
Routledge & Kegan Paul Ltd
Broadway House, 68–74 Carter Lane
London, E.C.4
and in the U.S.A. by
Dover Publications Inc.
180 Varick Street
New York, 10014
© D. O. Tall, 1970
No part of this book may be reproduced
in any form without permission from
the publisher, except for the quotation
of brief passages in criticism

ISBN 0 7100 6567 1 (p)
ISBN 0 7100 6850 6 (c)
Dover SBN 0 486 62663 6

Printed in Great Britain
by Willmer Brothers Limited,
Birkenhead

Preface

Functions of a Complex Variable I, II together form a sequel to *Complex Numbers* by Walter Ledermann. They contain an elementary introduction to complex differential and integral calculus.

The present text is mainly concerned with differential calculus. However certain results in the theory (for example Taylor's series) are best proved by using integral calculus and so contour integration is introduced at an early stage. The complex analogue of the Fundamental Theorem of Calculus (which exhibits integration and differentiation as inverse operations) is discussed and this gives a natural approach to Cauchy's Theorem. This in turn yields a proof of Taylor's Theorem and a number of remarkable corollaries. For example one result states that if a complex function is assumed differentiable just once everywhere in its domain of definition, then all the higher derivatives automatically exist.

Volume II contains applications of the theory, including conformal mappings, harmonic functions, calculation of integrals by residues, together with a description of analytic continuation and Riemann surfaces.

The main difficulties encountered in the text concern the theory of curves, since the sophisticated techniques (compactness, winding number etc.) which are necessary to deal with arbitrary curves are too technical for a book of this nature. Where difficulties occur, they are clearly stated. The general theory is illustrated by examples and each chapter ends with a set of exercises for the reader.

I should like to thank my colleagues Professor Walter Ledermann and Dr. Alan Weir for reading the manuscript and making many helpful suggestions and improvements in the text.

The University of Warwick DAVID TALL

Contents

CHAPTER ONE

Differentiation

1. Preliminaries

We begin with an informal discussion on functions as a prelude to more precise assumptions which will be explained in the next section.

The notion of a function of a complex variable is intuitively very clear. Given a complex number z, there is defined uniquely another complex number $f(z)$. This is usually given by a formula, for example $f(z) = z^2$ or $f(z) = e^z$.

We also wish to consider such formulae as $f(z) = 1/z$ or the power series $f(z) = 1+z+z^2 \ldots$ to give functions. These differ from the preceding examples in that they are not defined for every value of z. The formula $f(z) = 1/z$ is not defined for $z = 0$ and the power series is not convergent (and hence the sum is not defined) for $|z|>1$. However they have in common with $f(z) = z^2$ and $f(z) = e^z$ the property that, for any given value of z, if $f(z)$ is defined then $f(z)$ is unique. This is not so for the expression $z^{\frac{1}{2}}$ which has two values for $z \neq 0$ or for $\log z$ which† has many values for $z \neq 0$. Such expressions are sometimes called 'many-valued functions' and they will be considered separately later. In the remainder of the text a function will always be assumed to be single-valued wherever it is defined.

If we write $z = x+iy$ where x, y are real and $f(z) = u+iv$ where u, v are real, then u, v are real functions of x, y. For this reason we write $f(z) = u(x, y)+iv(x, y)$ to illustrate that u, v depend on x, y.

† W. Ledermann, *Complex Numbers*, in this series, p. 57.

EXAMPLE 1. If $f(z) = z^2$, then $f(z) = (x+iy)^2$ $= x^2-y^2+2ixy$ and so $u(x, y) = x^2-y^2$, $v(x, y) = 2xy$.

EXAMPLE 2. If $f(z) = e^z$, then† $f(z) = e^{x+iy}$ $= e^x(\cos y+i \sin y)$ and so $u(x, y) = e^x \cos y$, $v(x, y) = e^x \sin y$.

2. The Domain of Definition of a Function

As we have seen, we wish to consider functions which are not defined everywhere. The set of complex numbers where a function is defined will be called the domain of definition. We wish to put certain restrictions on this set and these ideas are discussed in this section. It is convenient to consider the situation pictorially by identifying complex numbers with points in the plane.

If z_0 is a complex number, the *ε-neighbourhood* of z_0 is the set of all points z such that $|z-z_0| < \varepsilon$ where ε is a given positive real number.

In figure 1, the ε-neighbourhood of z_0 is the set of points in the shaded disc not including the boundary.

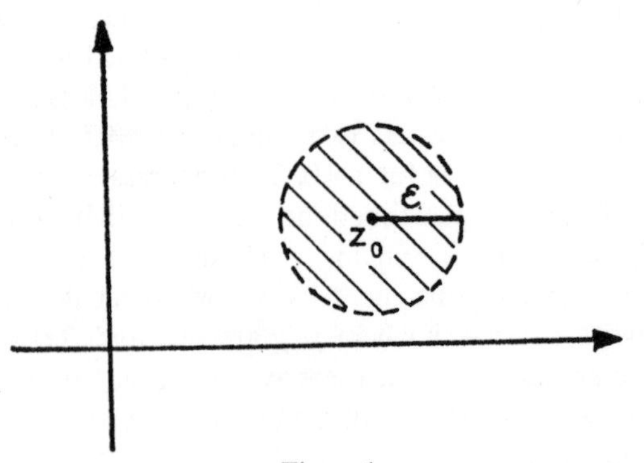

Figure 1

† W. Ledermann, *Complex Numbers*, in this series, p. 56.

A set S of points in the complex plane is said to be *open* if every point z_0 in S has an ε-neighbournood which consists entirely of points of S. For example the set C of points such that $|z| < 1$ is open, for if z_0 is in C, let $|z_0| = 1 - \delta$ where $0 < \delta \leqslant 1$, then the ε-neighbourhood of z_0 where $0 < \varepsilon \leqslant \delta$ lies completely in C.

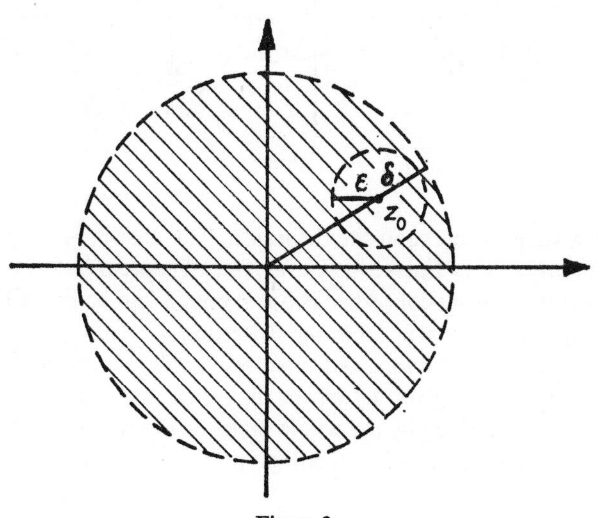

Figure 2

A *stepwise curve* in the plane is a polygonal curve, all of whose straight segments are parallel either to the real or imaginary axis.

A set S of points in the complex plane is said to be *connected* if any two points in S may be joined by a stepwise curve which lies entirely in S.

The shaded area in figure 3 is connected where z_1, z_2 are typical points.

Remark. The reader is perfectly justified in asking why we

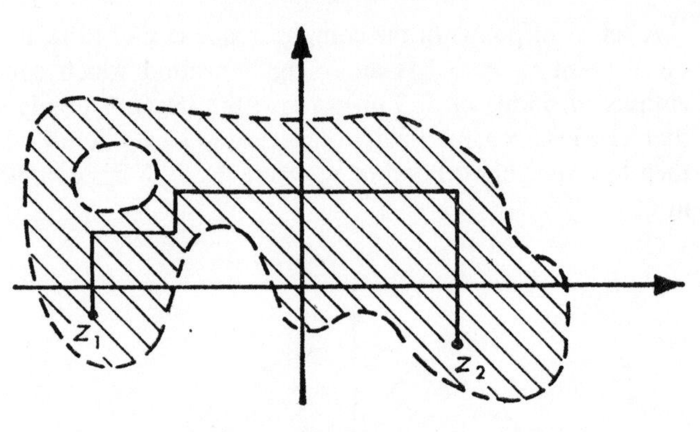

Figure 3

use a stepwise curve in the definition. The answer is simple; it is because it is the most useful (see Theorem 5.1. below). Actually if the set concerned is open, then it can be proved that any type of curve will do in the definition.

A (non-empty) connected open set is called a *domain*†. For example the set given by $|z| < 1$ (figure 2) is a domain. Other examples are given by the whole plane or the whole plane with a finite number of points missing.

Fundamental assumption. A complex function will always be assumed to be defined on a domain. This is called the *domain of definition* of the function concerned. Thus a *function of a complex variable* will be a rule which assigns to each complex number z in the domain of definition a unique complex number $f(z)$.

This rule is usually given by a 'formula' such as e^z or $1/z$ and, in common with the usual practice, we will often refer to the function by this formula.

† Some texts use the word 'region' instead of 'domain'.

Examples of functions are

(i) e^z, defined for all z,
(ii) $1/z$, defined for $z \neq 0$,
(iii) $1+z+z^2+\ldots$, defined for $|z|<1$.

As we have remarked, log z is *not* a function in the sense that it is not single-valued. We recall that

$$\log z = \log |z| + i\,(\arg z + 2\pi k)$$

where $\log |z|$ is the usual real logarithm, $-\pi < \arg z \leqslant \pi$ and k is an integer‡. We can consider log z to be a function in the following manner: let the 'cut-plane' consist of the complex plane with the negative real axis (including zero) removed:

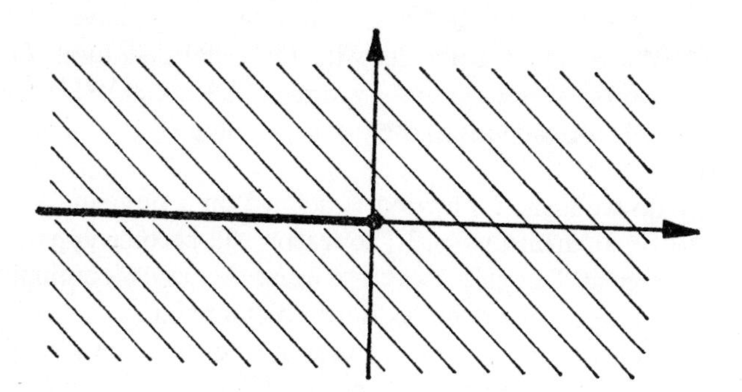

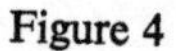

Figure 4

Now choose a fixed value of k and then log z is a single-valued function in the cut-plane. For example the principal value given by $k = 0$,

$$\text{Log } z = \log |z| + i \arg z$$

where now $-\pi < \arg z < \pi$.

‡ W. Ledermann, *Complex Numbers*, p. 57.

Note that the cut-plane is a domain, so now $\log z$ is a function according to our definition.

In a similar manner we can consider $z^{\frac{1}{2}}$ to be a function in the cut-plane. Write $z = re^{i\theta}$ where $-\pi < \theta < \pi$ in the cut-plane, then choose the value $z^{\frac{1}{2}} = r^{\frac{1}{2}}e^{\frac{1}{2}i\theta}$. This is a function and it is usually referred to as the principal value. On the positive real axis where $\theta = 0$, it reduces to the positive square root $r^{\frac{1}{2}}$. The other value $z^{\frac{1}{2}} = r^{\frac{1}{2}}e^{i(\frac{1}{2}\theta+\pi)}$ is also a function in the cut-plane. On the positive real axis, it reduces to $r^{\frac{1}{2}}e^{i\pi} = -r^{\frac{1}{2}}$, the negative square root.

As a further example of a function defined in the cut-plane, we define the principal value of z^{α} to be $e^{\alpha \operatorname{Log} z}$ where α is any complex constant. Since $\operatorname{Log} z$ is uniquely defined in the cut-plane, z^{α} is well-defined, for example the principal value of $i^{i} = e^{i \operatorname{Log} i} = e^{i.i(\pi/2)} = e^{-\pi/2}$. For an integer n we have $e^{n \operatorname{Log} z} = (e^{\operatorname{Log} z})^{n} = z^{n}$ which coincides with the usual definition. For $\alpha = \frac{1}{2}$, $z = re^{i\theta}$ where $-\pi < \theta < \pi$, then $e^{\frac{1}{2}\operatorname{Log} z} = e^{\frac{1}{2}\operatorname{Log} r + \frac{1}{2}i\theta} = r^{\frac{1}{2}}e^{\frac{1}{2}i\theta}$ which corresponds to the principal value of $z^{\frac{1}{2}}$ as defined above.

Why do we insist a function is defined on a domain? Why not just on an arbitrary set? The reason will become apparent as we progress. Roughly speaking, when we discuss continuity or differentiation of a complex function at a point z_0, we would like the function to be defined near z_0 (i.e. in some ε-neighbourhood) and so require the region of definition to be open. The reason for connectedness is more subtle. It pays dividends when the function concerned is differentiable. If the function were defined on a set which consisted of several disjoint parts, the function could 'behave quite differently' on each piece. For example we could have $f(z) = z$ for $|z| < 1$, $f(z) = e^{z}$ for $|z| > 2$ and not defined for $1 \leqslant |z| \leqslant 2$. However if the set where the function is defined is a domain (in particular connected) and the function is differentiable, then this imposes quite strict conditions on it. Indeed, it can be shown that if we know the values of the function on part of the domain, it is determined everywhere! We leave a precise statement and proof of this remarkable result until Chapter III, but mention it to justify the introduction of a "domain of definition".

Pictorially it is impossible to represent a complex function completely and we cannot draw a graph. This is because a complex number is represented as a point in two-dimensional real space. So we would need two dimensions to represent the values of z and two for $f(z)$, making a total of four. Since we have only two-dimensional paper at our disposal, the best we can do is imagine two complex planes and as z moves about in the first, $f(z)$ moves about in the second. Of course the only values of z for which $f(z)$ is defined are those in the domain of definition so we could illustrate this by drawing the domain of definition in the first plane (denoted by D and shaded):

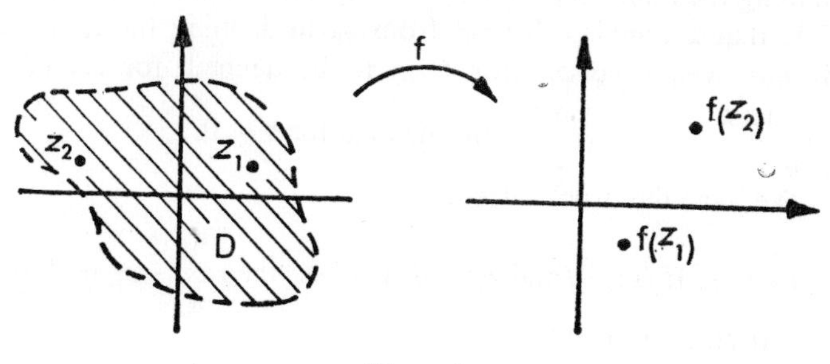

Figure 5

3. Limits and Continuity

The definitions of these concepts for complex functions are the same as in the real case†. Hence the reader familiar with the real case should find no difficulty.

DEFINITION 3.1. We say that $f(z)$ tends to the limit l as z

† P. J. Hilton, *Differential Calculus*, this series, pp. 10, 11.

tends to z_0 if the distance from $f(z)$ to l remains as small as we please so long as z remains sufficiently near to z_0, while remaining distinct from z_0.

We write $f(z) \to l$ as $z \to z_0$ or $\lim_{z \to z_0} f(z) = l$. Of course the distance from $f(z)$ to l is $|f(z) - l|$ and saying z is sufficiently near to z_0 means that $|z - z_0|$ is sufficiently small. So definition 3.1 could be phrased in terms of real numbers by defining $f(z) \to l$ as $z \to z_0$ to mean $|f(z) - l| \to 0$ as $|z - z_0| \to 0$. In precise terms, given $\varepsilon > 0$ (no matter how small), we can always find $\delta > 0$ (where δ may depend on ε) such that $0 < |z - z_0| < \delta$ implies $|f(z) - l| < \varepsilon$.

We make the usual remark that $\lim_{z \to z_0} f(z) = l$ does not mean the same as $f(z_0) = l$. The value of $f(z_0)$ is irrelevant in determining the limit because we have expressly stated in definition 3.1. that z remains distinct from z_0 in defining the limit. It is not even necessary for $f(z_0)$ to be defined, for example

$\lim_{z \to 0} \dfrac{\sin z}{z} = 1$ but $\dfrac{\sin z}{z}$ is not defined for $z = 0$.

We have the usual rules for limits:

RULES. If $f(z) \to l$ and $g(z) \to k$ as $z \to z_0$, then as $z \to z_0$ we have

(i) $f(z) + g(z) \to l + k$,
(ii) $f(z) - g(z) \to l - k$,
(iii) $f(z)g(z) \to lk$,
(iv) if $k \neq 0$, $f(z)/g(z) \to l/k$.

These results can either be proved from first principles as in the real case, or by resolving each complex number into its real and imaginary parts and arguing as for limits of sequences†.

DEFINITION 3.2. We say $f(z)$ is continuous at z_0 if $f(z_0)$ is defined and $\lim_{z \to z_0} f(z)$ exists and equals $f(z_0)$.

† W. Ledermann, *Complex Numbers*, p. 47.

For example $f(z) = |z|$ is continuous everywhere. This is because $0 \leqslant ||z| - |z_0|| \leqslant |z - z_0|$, so if z is close to z_0, $|z - z_0|$ is small and $|f(z) - f(z_0)| = ||z| - |z_0||$ is small or even smaller. This shows that $f(z) \to f(z_0)$ as $z \to z_0$.

If we have a continuous function, we can imagine the situation pictorially, for if we consider a point z which approaches z_0, then the image $f(z)$ approaches $f(z_0)$.

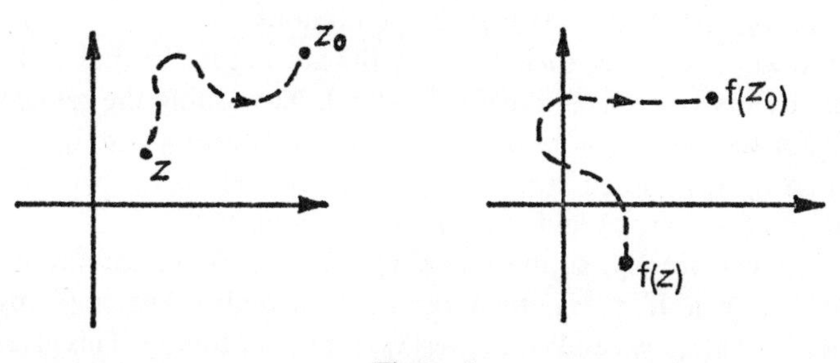

Figure 6

Continuity of a complex function is no more involved than the real case. Using rule (i) for limits, we see that if $f(z)$ and $g(z)$ are continuous at z_0, then as $z \to z_0$, we have $f(z) + g(z) \to f(z_0) + g(z_0)$ showing $f(z) + g(z)$ is continuous at z_0. Similarly the difference, product and quotient of continuous functions are continuous.

Also if $g(z)$ is continuous at z_0 and $f(w)$ is continuous at $w_0 = g(z_0)$, then $f(g(z))$ is continuous at z_0. This is because $z \to z_0$ implies $g(z) \to g(z_0) = w_0$ and so $f(g(z)) \to f(w_0) = f(g(z_0))$.

We can reduce the theory of continuity of a complex function to continuity of real functions of two real variables. Suppose $w = u + iv$, where u, v are real, then we first observe that $w \to w_0$ is equivalent to $u \to u_0$, $v \to v_0$ both together. To see this, note that

$$0 \leqslant |u - u_0| \leqslant \sqrt{\{(u - u_0)^2 + (v - v_0)^2\}} = |w - w_0|$$

and so $|u - u_0|$ is never greater than $|w - w_0|$. If $w \to w_0$ then $|w - w_0| \to 0$ implying $|u - u_0| \to 0$ and so $u \to u_0$. Similarly $v \to v_0$.

Conversely if both $u \to u_0$ and $v \to v_0$, then

$$|w - w_0| = \sqrt{\{(u-u_0)^2 + (v-v_0)^2\}} \to 0$$

and so $w \to w_0$. Using this fact we may prove:

THEOREM 3.1. If $f(z) = u(x, y) + iv(x, y)$ then the complex function $f(z)$ is continuous at $z_0 = x_0 + iy_0$ if and only if the real functions $u(x, y)$, $v(x, y)$ are continuous at (x_0, y_0).

Proof. (i) Suppose $f(z)$ is continuous at $z_0 = x_0 + iy_0$. If $x \to x_0$ and $y \to y_0$ then $z \to z_0$ by the above remark and so by continuity of $f(z)$, we have $f(z) \to f(z_0)$. Now apply the remark again to $w = f(z) = u(x, y) + iv(x, y)$ then $w \to w_0 = u(x_0, y_0) + iv(x_0, y_0)$ and so $u(x, y) \to u(x_0, y_0)$, $v(x, y) \to v(x_0, y_0)$. This shows that $u(x, y)$ and $v(x, y)$ are continuous.

(ii) Conversely, suppose $u(x, y)$ and $v(x, y)$ are continuous at (x_0, y_0). If $z \to z_0$, then both $x \to x_0$ and $y \to y_0$ implying $u(x, y) \to u(x_0, y_0)$ and $v(x, y) \to v(x_0, y_0)$ by continuity. This gives

$$u(x, y) + iv(x, y) \to u(x_0, y_0) + iv(x_0, y_0)$$

that is to say $f(z) \to f(z_0)$ and so $f(z)$ is continuous.

EXAMPLE 1. arg z is continuous in the cut-plane.

This is a basic result that we will need later and it is quite tricky to prove. The method we give uses a theorem from real variable theory†.

Suppose that $t = h(\theta)$ $(\alpha \leqslant \theta \leqslant \beta)$ is a real-valued monotonic strictly increasing function where $h(\alpha) = a$, $h(\beta) = b$, then we may solve this to find θ in terms of t, $\theta = g(t)$ $(a \leqslant t \leqslant b)$. Furthermore, if h is continuous, then so is g. For example $t = \sin\theta$ $\left(-\dfrac{\pi}{2} \leqslant \theta \leqslant \dfrac{\pi}{2}\right)$ is such a function, taking every value in $-1 \leqslant t \leqslant 1$ and so $\theta = \sin^{-1} t$ is well-defined and continuous

† Scott & Tims, *Mathematical Analysis*, Cambridge University Press, p. 217.

for $-1 \leqslant t \leqslant 1$, taking values in $-\frac{\pi}{2} \leqslant \sin^{-1} t \leqslant \frac{\pi}{2}$.

First consider the domain given by $x > 0$.

Here $\arg z = \sin^{-1}\left(\frac{y}{\sqrt{(x^2+y^2)}}\right)$

where we choose $-\frac{\pi}{2} < \sin^{-1}\left(\frac{y}{\sqrt{(x^2+y^2)}}\right) < \frac{\pi}{2}$.

But $\sqrt{(x^2+y^2)} = |z|$ is continuous and non-zero for $x > 0$, hence $y/|z|$ is a continuous function of x, y for $x > 0$. Thus $\sin^{-1}(y/|z|)$ is a continuous function of a continuous function and hence continuous. This shows that $\arg z$ is continuous in the domain $x > 0$.

Similarly we may show that $\arg z$ is continuous for $y > 0$ by considering $\arg z = \cos^{-1}\left(\frac{y}{\sqrt{(x^2+y^2)}}\right)$ where we choose $0 \leqslant \cos^{-1}\left(\frac{y}{\sqrt{(x^2+y^2)}}\right) \leqslant \pi$. (Note that cos is monotonic decreasing here.) Finally $\arg z$ is continuous for $y < 0$ where we have $\arg z = \cos^{-1}\left(\frac{x}{\sqrt{(x^2+y^2)}}\right)$, this time choosing $-\pi \leqslant \cos^{-1}\left(\frac{x}{\sqrt{(x^2+y^2)}}\right) \leqslant 0$. The three domains $x > 0$, $y > 0$, $y < 0$ together cover the cut-plane and the result is proved.

EXAMPLE 2. Log z is continuous in the cut-plane.
This follows from example 1 and theorem 3.1 because $\text{Log}\, z = \log|z| + i \arg z$. Note that the real part $\log|z| = \log(x^2+y^2)^{\frac{1}{2}}$ is continuous for $(x, y) \neq (0, 0)$ and the imaginary part is continuous in the cut-plane. This gives the required result.

4. Differentiation

As with limits and continuity, differentiation of a complex

function is defined in the same way as the real case.† (Notice however that the derivative is no longer the gradient of a graph because we cannot draw the graph of a complex function.)

DEFINITION 4.1. The derivative at z_0 of the function $f(z)$ is

$$f'(z_0) = \lim_{z \to z_0} \frac{f(z)-f(z_0)}{z-z_0}.$$

If we make a change of variable $z-z_0 = h$ we also have

$$f'(z_0) = \lim_{h \to 0} \frac{f(z_0+h)-f(z_0)}{h}.$$

EXAMPLE. $f(z) = z^2$

$$f'(z_0) = \lim_{h \to 0} \frac{(z_0+h)^2-{z_0}^2}{h} = \lim_{h \to 0} (2z_0+h) = 2z_0 .$$

Often we use the alternative notation $w = f(z)$ and $\frac{dw}{dz} = f'(z)$.

Of course $f'(z_0)$ may not exist. As a trivial example, if $f(0) = 0$ and $f(z) = 1$ for $z \neq 0$, then $f'(0)$ does not exist. For $h \neq 0$, $\frac{f(h)-f(0)}{h} = \frac{1}{h}$ and this does not tend to a finite limit as $h \to 0$. We now show that a differentiable function is necessarily continuous (thus we may infer that a discontinuous function cannot be differentiable).

THEOREM 4.1. If $f(z)$ is differentiable at z_0, then $f(z)$ is continuous at z_0.

Proof. $\lim_{z \to z_0} (f(z)-f(z_0)) = \lim_{z \to z_0} \frac{(f(z)-f(z_0))}{z-z_0}(z-z_0)$

$= \lim_{z \to z_0} \frac{f(z)-f(z_0)}{z-z_0} \lim_{z \to z_0} (z-z_0)$ by the rule for limits

† P. J. Hilton, *Differential Calculus*, p. 12.

$= f'(z_0).0$

$= 0.$

So $\lim_{z \to z_0} f(z) = f(z_0)$ and $f(z)$ is continuous at z_0.

We may verify the usual rules for differentiation as in the real case†:

RULES.

(i) $\frac{d}{dz}(Af(z)+Bg(z)) = Af'(z)+Bg'(z)$ where A, B are (complex) constants.

(ii) $\frac{d}{dz}(f(z)\,g(z)) = f(z)\,g'(z)+f'(z)\,g(z)$

(iii) $\frac{d}{dz}(f(z)/g(z)) = \{f'(z)\,g(z)-f(z)\,g'(z)\}/(g(z))^2$ if $g(z) \neq 0$

(iv) $\frac{d}{dz}f(g(z)) = f'(g(z))\,g'(z).$

EXAMPLE. If n is an integer, $\frac{d}{dz}(z^n) = nz^{n-1}$.

Proof by induction. For $n = 1$, $\frac{d}{dz}(z) = \lim_{h \to 0} \frac{(z+h)-z}{h} = 1$ and so the formula is true.

Assume it for n, then $\frac{d}{dz}(z^{n+1}) = \frac{d}{dz}(z^n z)$

$= z^n \frac{d}{dz}(z) + \frac{d}{dz}(z^n)z$ by (ii)

$= z^n + nz^{n-1}z$

$= (n+1)z^n$ and so the formula is true for $n+1$ and by induction true for all positive integers.

Using (i) and (iv), we may calculate the derivative of a rational

† P. J. Hilton, *Differential Calculus*, pp. 16–19.

function $\dfrac{a_n z^n + \ldots\ldots + a_0}{b_m z^m + \ldots\ldots + b_0}$ in the same way, for example

$$\frac{d}{dz}\frac{z+1}{z^2+2} = \left\{\frac{d}{dz}(z+1)\,.\,(z^2+2)-(z+1)\frac{d}{dz}(z^2+2)\right\}\Big/(z^2+2)^2$$
$$= (z^2+2-2z^2-2z)/(z^2+2)^2$$
$$= (2-2z-z^2)/(z^2+2)^2$$

A rational function is differentiable wherever it is defined (i.e. whenever $b_m z^m + \ldots\ldots + b_0 \neq 0$).

DEFINITION 4.2. A function of a complex variable is said to be *analytic*† if it is differentiable everywhere in its domain of definition.

For example rational functions are analytic.

5. The Cauchy-Riemann Equations

We now come across the first property that distinguishes the complex theory from the real. When calculating $f'(z_0) = \lim\limits_{z\to z_0}\dfrac{f(z_0)-f(z)}{z-z_0}$, we may let z approach z_0 in *any* fashion.

Let us calculate $f'(z_0)$ in two distinct ways:

(i) Let $z_0 = x_0+iy_0$, $z = x_0+h+iy_0$ where h is real, and write $f(z) = u(x, y)+iv(x, y)$ then $f'(z_0) = \lim\limits_{z\to z_0}\dfrac{f(z)-f(z_0)}{z-z_0}$

$$= \lim_{h\to 0}\left\{\frac{f((x_0+h)+iy_0)-f(x_0+iy_0)}{h}\right\}$$

$$= \lim_{h\to 0}\left\{\frac{u(x_0+h, y_0)+iv(x_0+h, y_0)-u(x_0, y_0)-iv(x_0, y_0)}{h}\right\}$$

$$= \lim_{h\to 0}\left\{\frac{u(x_0+h, y_0)-u(x_0, y_0)}{h}+\frac{i(v(x_0+h, y_0)-v(x_0, y_0))}{h}\right\}$$

† Some texts use the word ‘regular’ instead of ‘analytic’.

$$= \frac{\partial u}{\partial x} + i\frac{\partial v}{\partial x}.$$

(ii) Let $z_0 = x_0 + iy_0$, $z = x_0 + i(y_0 + k)$ where k is real, then as $z \to z_0$, we have $k \to 0$ and so

$$f'(z_0)$$

$$= \lim_{z \to z_0} \frac{f(z) - f(z_0)}{z - z_0}$$

$$= \lim_{k \to 0} \frac{f(x_0 + i(y_0 + k)) - f(x_0 + iy_0)}{ik}$$

$$= \lim_{k \to 0} \left\{ \frac{u(x_0, y_0 + k) + iv(x_0, y_0 + k) - u(x_0, y_0) - iv(x_0, y_0)}{ik} \right\}$$

$$= \lim_{k \to 0} \left\{ \frac{v(x_0, y_0 + k) - v(x_0, y_0)}{k} - \frac{i(u(x_0, y_0 + k) - u(x_0, y_0))}{k} \right\}$$

$$= \frac{\partial v}{\partial y} - i\frac{\partial u}{\partial y}.$$

Since $f'(z_0)$ is uniquely defined no matter how we let z approach z_0, we must have

$$f'(z_0) = \frac{\partial u}{\partial x} + i\frac{\partial v}{\partial x} = \frac{\partial v}{\partial y} - i\frac{\partial u}{\partial y}.$$

Comparing real and imaginary parts, we find that

$$\frac{\partial u}{\partial x} = \frac{\partial v}{\partial y}, \quad \frac{\partial v}{\partial x} = -\frac{\partial u}{\partial y}.$$

These are called the Cauchy-Riemann equations which hold for differentiable complex functions. They give a simple way of asserting a function is *not* differentiable.

EXAMPLE. $f(z) = |z|$. Here $u(x, y) = \sqrt{(x^2 + y^2)}$, $v(x, y) = 0$, giving:

$$\frac{\partial u}{\partial x} = \frac{2x}{\sqrt{(x^2 + y^2)}}, \quad \frac{\partial u}{\partial y} = \frac{2y}{\sqrt{(x^2 + y^2)}}, \quad \frac{\partial v}{\partial x} = 0 = \frac{\partial v}{\partial y}.$$

Hence if either $x \neq 0$ or $y \neq 0$, at least one of the equations $\frac{\partial u}{\partial x} = \frac{\partial v}{\partial y}$, $\frac{\partial v}{\partial x} = -\frac{\partial u}{\partial y}$ does not hold. If both $x = 0$ and $y = 0$, then substituting in $\frac{\partial u}{\partial x}, \frac{\partial u}{\partial y}$ we get $\frac{0}{0}$. Returning to first principles

$$\frac{\partial u}{\partial x}(0, 0) = \lim_{k \to 0} \frac{u(k, 0) - u(0, 0)}{k} = \lim_{k \to 0} \frac{\sqrt{k^2}}{k}$$

$$= \lim_{k \to 0} \frac{|k|}{k}.$$

Since

$$\frac{|k|}{k} = \begin{cases} 1 \text{ for } k > 0 \\ -1 \text{ for } k < 0, \end{cases}$$

the limit of $\frac{|k|}{k}$ as $k \to 0$ does not exist and so $\frac{\partial u}{\partial x}(0, 0)$ is not defined. Similarly $\frac{\partial v}{\partial x}(0, 0)$ does not exist and the Cauchy-Riemann equations cannot hold at the origin.

Thus $f(z) = |z|$ is not differentiable anywhere but it is continuous everywhere!

The reader who may be upset by this seemingly unnatural state of affairs may be consoled by the fact that as well as z^n, the standard functions, e^z, cos z, sin z etc. are all analytic. As a possible proof of this fact we might use the Cauchy-Riemann equations. This meets with a small obstacle.

If $f(z) = u(x, y) + iv(x, y)$ is a complex function such that the partial derivatives $\frac{\partial u}{\partial x}, \frac{\partial v}{\partial x}, \frac{\partial u}{\partial y}, \frac{\partial v}{\partial y}$ exist and satisfy the Cauchy-Riemann equations, then $f(z)$ need not be differentiable. For example the (rather synthetic) function given by $f(z) = 1$ if $x = 0$ or $y = 0$, but $f(z) = 0$ otherwise satisfies the Cauchy-Riemann equations at the origin (all partial derivatives are zero) but it is not differentiable there because it is not even continuous.

However if the partial derivatives $\frac{\partial u}{\partial x}, \frac{\partial v}{\partial x}, \frac{\partial u}{\partial y}, \frac{\partial v}{\partial y}$ exist, satisfy the Cauchy-Riemann equations and are all *continuous*, then we *can* infer that the function is differentiable. The proof of this fact will be omitted since we will not use the result later. However, as an example of its possible use, consider $f(z) = e^z = e^{x+iy} = e^x \cos y + ie^x \sin y$.

$$\frac{\partial u}{\partial x} = e^x \cos y = \frac{\partial v}{\partial y}, \quad \frac{\partial v}{\partial x} = e^x \sin y = -\frac{\partial u}{\partial y}$$

and all the partial derivatives are continuous, so by the above remark, e^z is differentiable with derivative

$$\frac{\partial u}{\partial x} + i\frac{\partial v}{\partial x} = \frac{\partial v}{\partial y} - i\frac{\partial u}{\partial y} = e^x \cos y + ie^x \sin y = e^z.$$

Thus we have verified the equation $\frac{d(e^z)}{dz} = e^z$, already well-known in its real form. In the next section we will demonstrate this in a different way using the power series expansion for e^z.

We close this section with the following:

THEOREM 5.1. If $f(z)$ is a function of a complex variable defined on a domain D, then $f'(z) = 0$ for all z in D implies $f(z)$ is constant.

Remark. If $f'(z) = 0$ and the function were defined on a set that was not connected, this theorem need not be true. The function could be constant on each connected piece, but the constants need not be the same. For example on the set $|z| < 1$ or $|z| > 2$, define $f(z) = 0$ for $|z| < 1$ and $f(z) = 1$ for $|z| > 2$ then $f'(z) = 0$, but $f(z)$ is not constant.

Proof of theorem. If $f'(z) = 0$ then

$$\frac{\partial u}{\partial x} + i\frac{\partial v}{\partial x} = \frac{\partial v}{\partial y} - i\frac{\partial u}{\partial y} = 0$$

implying

$$\frac{\partial u}{\partial x} = \frac{\partial v}{\partial x} = \frac{\partial u}{\partial y} = \frac{\partial v}{\partial y} = 0.$$

Now $\frac{\partial u}{\partial x} = 0$ i.e. $\frac{\partial u}{\partial x}(x, y_0) = 0$ for fixed y_0. From real variable theory† we have $u(x, y_0) =$ constant (since $u(x, y_0)$ is a real function of the real variable x). This means $u(x, y)$ is constant along any horizontal line segment $y = y_0$ (= constant) in the domain of definition. Similarly $\frac{\partial u}{\partial y} = 0$ implies that $u(x, y)$ is constant along any verticle line segment $x = x_0$ (= constant) in the domain of definition. Also from $\frac{\partial v}{\partial x} = \frac{\partial v}{\partial y} = 0$ we reach the same conclusions for $v(x, y)$. Hence $f(z) = u(x, y)+iv(x, y)$ is constant along each horizontal or vertical segment in the domain of definition.

But a domain is *connected*, (here comes the full force of the definition), and any two points z_1, z_2 in the domain of definition may be joined by a stepwise curve which lies entirely in the domain (refer back to figure 3). Since $f(z)$ is constant along each segment of this curve, we must have $f(z_1) = f(z_2)$. Since z_1, z_2 are arbitrary points in the domain, $f(z)$ is constant.

6. Power Series

We assume the reader has already met the idea of a power series

$$c_0+c_1z+c_2z^2+ \ldots\ldots +c_nz^n+ \ldots\ldots$$

where z is a complex variable and $c_0, c_1, c_2, \ldots$ are fixed complex numbers‡. We recall that either there is a positive number R (called the radius of convergence) such that the series

† P. J. Hilton, *Differential Calculus*, p. 37.
‡ W. Ledermann, *Complex Numbers*, p. 49.

converges absolutely for $|z| < R$ and diverges for $|z| > R$, or the series converges absolutely for all z (in which case we formally write $R = \infty$)†. Thus in our terminology the power series is a function with domain $|z| < R$.

EXAMPLE (i). $1+z+z^2+\ldots\ldots+z^n+\ldots\ldots$ has radius of convergence $R = 1$, and for $|z| < 1$, we have

$$1+z+z^2+\ldots\ldots+z^n+\ldots\ldots = (1-z)^{-1}.$$

EXAMPLES (ii)—(iv) all have infinite radius of convergence.

(ii) $e^z = 1+\frac{z}{1!}+\frac{z^2}{2!}+\ldots\ldots+\frac{z^n}{n!}+\ldots\ldots$

(iii) $\sin z = z-\frac{z^3}{3!}+\frac{z^5}{5!}-\ldots\ldots$

(iv) $\cos z = 1-\frac{z^2}{2!}+\frac{z^4}{4!}-\ldots\ldots$

It has already been noted that inside the circle of convergence power series may be manipulated in much the same way as polynomials‡. For example two power series may be added term by term. The same is true of differentiation. A power series may be differentiated term by term inside the circle of convergence and if $f(z) = c_0+c_1z+c_2z^2+\ldots+c_nz^n+\ldots$ for $|z| < R$, then $f'(z) = c_1+2c_2z+\ldots+nc_nz^{n-1}+\ldots$ for $|z| < R$. A proof of this result is somewhat technical and may be found in Appendix I. Note that a power series is differentiable everywhere in its domain of definition and so it is an analytic function.

EXAMPLE (i). $f(z) = 1+z+z^2+\ldots+z^n+\ldots \quad |z| < 1$, then $f'(z) = 1+2z+\ldots\ldots+nz^{n-1}+\ldots \quad |z| < 1$.

† W. Ledermann, *Complex Numbers*, p. 50.
‡ W. Ledermann, *Complex Numbers*, p. 51.

Since $f(z) = (1-z)^{-1}$ in this case, differentiating we have $f'(z) = (1-z)^{-2}$ and so

$$(1-z)^{-2} = 1+2z+ \dots +nz^{n-1}+ \dots \; |z|<1.$$

$$\text{(ii)}\; e^z = 1+\frac{z}{1!}+\frac{z^2}{2!}+ \dots +\frac{z^n}{n!}+ \dots$$

$$\frac{d}{dz}(e^z) = e^z.$$

$$\text{(iii)}\; \sin z = z-\frac{z^3}{3!}+\frac{z^5}{5!}- \dots\dots$$

$$\frac{d}{dz}(\sin z) = \cos z.$$

Similarly (iv) $\dfrac{d}{dz}(\cos z) = -\sin z.$

EXAMPLE (v). Since $w = \operatorname{Log} z$ is defined by the equation $z = e^w$, we can use this to show $\dfrac{dw}{dz} = \dfrac{1}{z}$. The function $w = \operatorname{Log} z$ is continuous in the cut-plane and so as $z \to z_0$, we have $w \to w_0$.

$$\begin{aligned}
\frac{dw}{dz} &= \lim_{z\to z_0} \frac{\operatorname{Log} z - \operatorname{Log} z_0}{z-z_0} \\
&= \lim_{w\to w_0} \frac{w-w_0}{e^w-e^{w_0}} \\
&= \lim_{w\to w_0} \left\{\frac{e^w-e^{w_0}}{w-w_0}\right\}^{-1} \\
&= 1 \Big/ \frac{d}{dw}(e^w) \\
&= 1/e^w \\
&= 1/z.
\end{aligned}$$

EXAMPLE (vi). $\frac{d}{dz}(z^\alpha) = \alpha z^{\alpha-1}$ in the cut-plane.

This is because $z^\alpha = e^{\alpha \operatorname{Log} z} = f(g(z))$ where $f(w) = e^w$, $g(z) = \alpha \operatorname{Log} z$ and so $\frac{d}{dz}(z^\alpha) = f'(g(z))g'(z) = e^{\alpha \operatorname{Log} z} \cdot \frac{\alpha}{z} = z^\alpha \cdot \frac{\alpha}{z} = \alpha z^{\alpha-1}$.

Notice that the derivative of a power series is again a power series with the same domain of definition. This means we can differentiate it again, indeed we can differentiate it as many times as we like, so that if

$$f(z) = c_0 + c_1 z + c_2 z^2 + c_3 z^3 + \ldots . + c_n z^n + \ldots .$$

then

$$f'(z) = c_1 + 2c_2 z + 3c_3 z^2 + 4c_4 z^3 + \ldots . + nc_n z^{n-1} + \ldots$$
$$f''(z) = 2c_2 + 6c_3 z + 12c_4 z^2 + \ldots + n(n-1)c_n z^{n-2} + \ldots$$
$$f'''(z) = 6c_3 + 24c_4 z + \ldots . + n(n-1)(n-2)c_n z^{n-3} + \ldots$$

etc.

Putting $z = 0$ in these equations we find $f(0) = c_0$, $f'(0) = c_1 = 1!c_1$, $f''(0) = 2c_2 = 2!c_2$, $f'''(0) = 6c_3 = 3!c_3$ and in general $f^{(n)}(0) = n!c_n$. This means that by substituting these values in the series we may write it in its usual Taylor-MacLaurin form:

$$f(z) = c_0 + c_1 z + c_2 z^2 + \ldots . + c_n z^n + \ldots .$$
$$= f(0) + \frac{f'(0)}{1!} z + \frac{f''(0)}{2!} z^2 + \ldots . + \frac{f^{(n)}(0)}{n!} z^n + \ldots .$$

More generally we may consider a power series centred on z_0:

$$f(z) = a_0 + a_1(z - z_0) + \ldots . + a_n(z - z_0)^n + \ldots \text{ for } |z - z_0| < R.$$

In this case we have:

$$f'(z) = a_1+2a_2(z-z_0)+ \dots +na_n(z-z_0)^{n-1}+ \dots |z-z_0|<R.$$

etc.

and we find $f^{(n)}(z_0) = n!a_n$.

This gives:

PROPOSITION 6.1. If $f(z) = a_0+a_1(z-z_0)+ \dots +$ $a_n(z-z_0)^n+ \dots$ for $|z-z_0|<R$, then f is differentiable as many times as we please for $|z-z_0|<R$ and $a_n = \dfrac{f^{(n)}(z_0)}{n!}$.

We remark at this stage that power series seem to be very special. They are not only differentiable, we can differentiate as many times as we please. In the real theory it is possible to invent functions which are differentiable once, but not twice. (e.g. $f(x) = 0$ for $x \leqslant 0$ and $f(x) = x^2$ for $x \geqslant 0$. Here $f'(x) = 0$ for $x \leqslant 0$ and $f'(x) = 2x$ for $x \geqslant 0$. Note that $f'(0) = 0$ calculated from either side. However $f''(0)$ does not exist, being 0 calculated from the left and 2 calculated from the right.)

It is a very pleasant (and surprising) fact that in the complex theory, if a function is analytic (i.e. differentiable once everywhere in its domain of definition) then it is differentiable as many times as we please. We will demonstrate this fact later (pages 55–56). How is it proved?—By using power series.

EXERCISES ON CHAPTER ONE

1. Write $f(z) = u(x, y)+iv(x, y)$ and find $u(x, y)$, $v(x, y)$ in each of the following cases:
   (i) z^2+2z (ii) $1/z$ $(z \neq 0)$ (iii) $\sin z$ (iv) $z/(e^z-1)$ $(z \neq 0)$
   (v) $\operatorname{Log} z$ in the cut-plane (vi) $|z|^2$ (vii) $\arg z$ in the cut-plane.
2. In exercise 1, differentiate (i)–(v).

3. (a) In exercise 1 (vi), calculate $\frac{\partial u}{\partial x}, \frac{\partial u}{\partial y}, \frac{\partial v}{\partial x}, \frac{\partial v}{\partial y}$ for $(x, y) \neq (0, 0)$. Hence show that $|z|^2$ is not differentiable for $z \neq 0$. What happens at $z = 0$? (Hint: use first principles).

   (b) In 1 (vii), calculate $\frac{\partial u}{\partial x}, \frac{\partial u}{\partial y}, \frac{\partial v}{\partial x}, \frac{\partial v}{\partial y}$ and hence show that $\arg z$ is not differentiable anywhere.

4. In each of the following cases, draw a sketch of the given set and say if it is a domain where z is subject to the given restriction: (i) $|z-1|<2$ (ii) $|z| \leqslant 1$ (iii) $x<-1$ or $x>1$ where $z = x+iy$ (iv) $z \neq t$ where t is real and $t \leqslant 0$ (v) $1<|z|<2$.

5. Suppose that f is an analytic function such that $f(z)$ is always real. Use the Cauchy-Riemann equations to prove that f is constant.

6. (i) Substitute $w = e^{\lambda z}$ in the equation $\frac{d^2w}{dz^2}+k^2w = 0$ $(k \neq 0)$ to find λ_1, λ_2 such that $e^{\lambda_1 z}$ and $e^{\lambda_2 z}$ are solutions. Show that $Ae^{\lambda_1 z}+Be^{\lambda_2 z}$ is also a solution where A, B are complex constants.

   Use the same method to find solutions for

   (ii) $\frac{d^2w}{dz^2}-\frac{3dw}{dz}+2w = 0$ (iii) $\frac{d^3w}{dz^3}-\frac{d^2w}{dz^2}+\frac{4dw}{dz}-4w = 0.$

7. Use $(1-z)^{-2} = 1+2z+3z^2+\ldots+nz^{n-1}+\ldots$ $|z|<1$ to find, by differentiation, a power series formula for $(1-z)^{-4}$ valid for $|z|<1$.

8. Consider $f(z) = 1+\sum_{n=1}^{\infty}\frac{\alpha(\alpha-1)\ldots\ldots(\alpha-n+1)}{n!}z^n$

   Prove that the series is absolutely convergent for $|z|<1$ and that $f'(z) = \alpha f(z)/(1+z)$. Consider $\phi(z) = \frac{f(z)}{(1+z)^\alpha}$ and show that $\phi'(z) = 0$ for $|z|<1$. Hence conclude that $f(z) = (1+z)^\alpha$ for $|z|<1$.

9. Show $f(z) = z-\frac{z^2}{2}+\frac{z^3}{3}-\frac{z^4}{4}+\ldots\ldots$ is absolutely convergent for $|z|<1$ and that $f'(z) = (1+z)^{-1}$.
   Hence conclude that $f(z) = \text{Log}(1+z)$ for $|z|<1$.

CHAPTER TWO

Integration

1. Contours

If $\phi(t)$, $\psi(t)$ are continuous real functions of the real variable t defined in an interval $\alpha \leqslant t \leqslant \beta$, the equation

$$z(t) = \phi(t)+i\psi(t) \quad (\alpha \leqslant t \leqslant \beta) \tag{1}$$

determines a *path* in the complex plane†. Thus a path is a continuous function defined on the interval $\alpha \leqslant t \leqslant \beta$, taking values in the complex plane. The *initial* and *final* points are $z(\alpha)$, $z(\beta)$ respectively and the path is said to be *closed* if $z(\alpha) = z(\beta)$. As t increases, the point $z(t)$ traverses a curve in the complex plane from $z(\alpha)$ to $z(\beta)$.

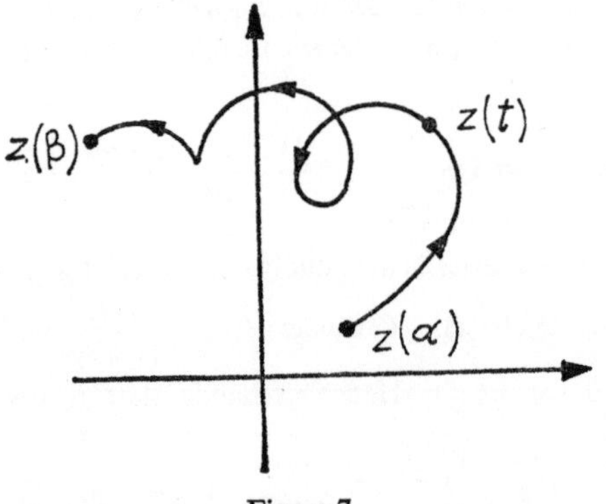

Figure 7

† This is the same definition as a path in the (x, y)-plane given in *Multiple Integrals* by W. Ledermann, p. 1.

The set of points in the complex plane given by $z(t)$ for $\alpha \leqslant t \leqslant \beta$ is called the *track*. Two different paths may have the same track, for example the paths

$$z_1(t) = \cos t + i \sin t \qquad \left(0 \leqslant t \leqslant \frac{\pi}{2}\right) \tag{2}$$

$$z_2(t) = \frac{1-t^2}{1+t^2} + \frac{2it}{1+t^2} \qquad (0 \leqslant t \leqslant 1) \tag{3}$$

both determine the track in figure 8:

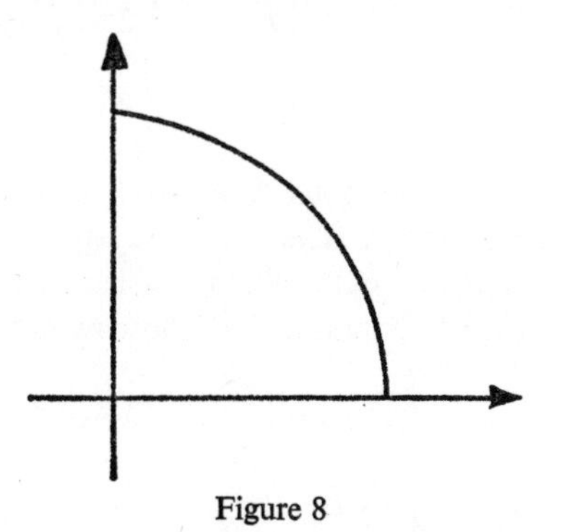

Figure 8

For this reason, if we refer to a pictorial representation of a curve and wish to talk about the path, then we should also specify the function which determines it. For example we will choose the standard function which represents the unit circle as a path to be

$$z(t) = \cos t + i \sin t \qquad (0 \leqslant t \leqslant 2\pi). \tag{4}$$

More generally, the circle centre z_0, radius r will be

$$z(t) = z_0 + re^{it} \qquad (0 \leqslant t \leqslant 2\pi). \tag{5}$$

The line segment from z_1 to z_2 will be

$$z(t) = z_1(1-t) + z_2 t \qquad (0 \leqslant t \leqslant 1). \tag{6}$$

We sometimes refer to the path as a 'parametrization' of the track and call t the 'parameter'.

The *opposite path* to (1) is the path

$$\begin{aligned} z_0(t) &= z(\alpha+\beta-t) \\ &= \phi(\alpha+\beta-t)+i\psi(\alpha+\beta-t) \qquad (\alpha \leqslant t \leqslant \beta). \end{aligned} \tag{7}$$

Notice that as t increases, $z_0(t)$ traverses the same track as $z(t)$, but in the opposite sense. For example the opposite path to (2) is

$$\begin{aligned} z_0(t) &= \cos\left(\frac{\pi}{2}-t\right)+i\sin\left(\frac{\pi}{2}-t\right) \\ &= \sin t+i\cos t \qquad \left(0 \leqslant t \leqslant \frac{\pi}{2}\right) \end{aligned} \tag{8}$$

A path is said to be *smooth* if $\phi'(t)$, $\psi'(t)$ exist and are continuous for $\alpha \leqslant t \leqslant \beta$. The paths (2)–(6) and (8) are all smooth. A *contour* is a path which consists of a finite number of smooth pieces. For example the following path is a contour:

$$z(t) = \begin{cases} t^2+i\sin\frac{\pi}{2}t & (0 \leqslant t \leqslant 1) \\ t+it & (1 \leqslant t \leqslant 2) \\ 4-t+2i & (2 \leqslant t \leqslant 3). \end{cases}$$

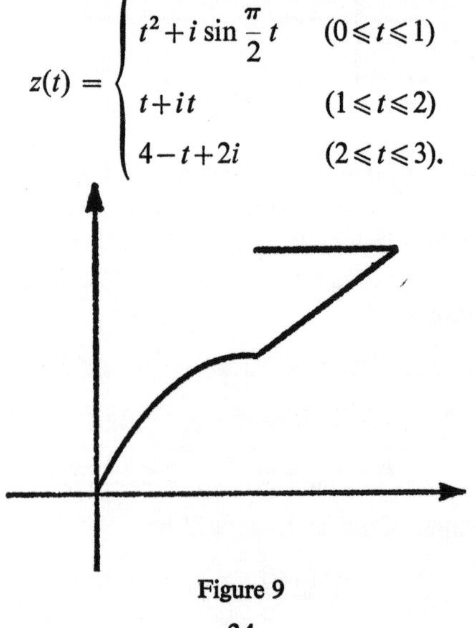

Figure 9

A *Jordan contour* is a contour

$$z(t) = \phi(t)+i\psi(t) \qquad (\alpha \leqslant t \leqslant \beta)$$

such that $t_1 \neq t_2$ implies $z(t_1) \neq z(t_2)$. Thus a Jordan contour has no self-intersections. A *closed Jordan contour* is a closed contour such that $\alpha \leqslant t_1 < t_2 < \beta$ implies $z(t_1) \neq z(t_2)$. In this case there are no self-intersections other than the coincident endpoints. An example is given by the unit circle

$$z(t) = \cos t + i \sin t \qquad (0 \leqslant t \leqslant 2\pi).$$

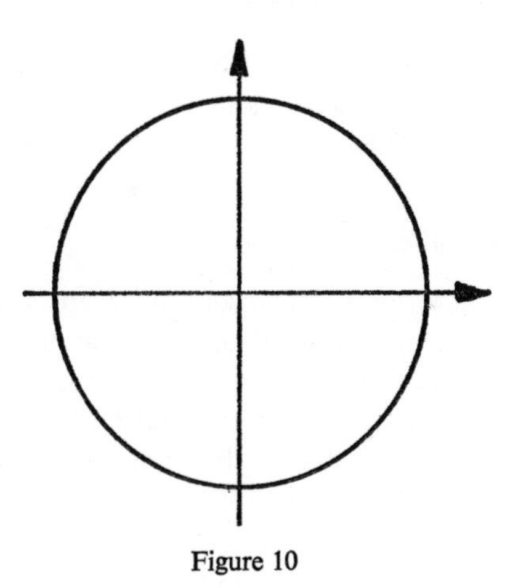

Figure 10

The *Jordan curve theorem* states that a closed Jordan contour divides the plane into two domains, one bounded (called the interior) and one unbounded (the exterior).

The reader who relies on his geometric intuition may feel that this result is patently obvious. For example the interior of the unit circle in figure 10 is certainly given by $|z| < 1$ and the exterior by $|z| > 1$. For every particular Jordan contour we meet in this text, the result will be clear. However it is possible to draw 'maze-like' Jordan curves such as:

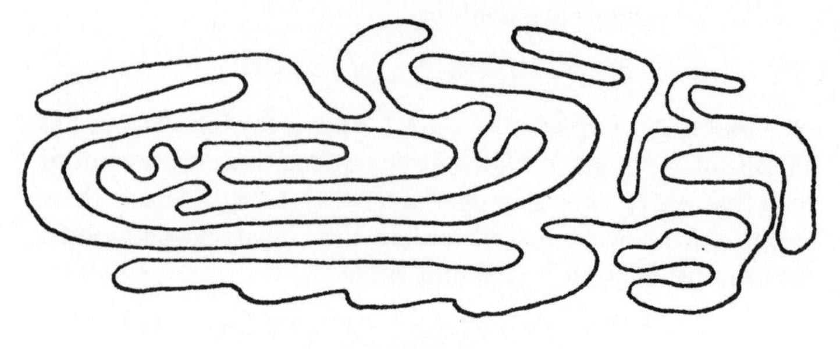

Figure 11

A proof of the theorem must be written so as to include every possiblity which may arise and it is not surprising that this is very difficult. For this reason the proof is omitted.

2. Contour Integration

As in the real case, we wish to discuss integration of a complex function. There is no immediate analogue to the symbol $\int_{z_1}^{z_2} f(z)\, dz$ where f is a complex function and z_1, z_2 are complex numbers. This is because we may consider z_1, z_2 as points in the plane and the symbol $\int_{z_1}^{z_2} f(z)\, dz$ does not specify how z varies between z_1 and z_2. To do this, the integral is defined along a contour γ from z_1 to z_2. In general this will depend on the choice of γ and so we use the notation $\int_\gamma f(z)dz$.

Note that we require f to be defined everywhere on the track of γ. Equivalently, if f is defined in the domain D, we could ask that γ lies in D (meaning, of course, that the track of γ lies in D). This is to be preferred, because then we can visualize a picture of the domain with (the track of) γ lying in it.

First we define an integral along a smooth path. Let f be a continuous complex function defined in a domain D and suppose that γ is the smooth path $z(t) = x(t)+iy(t)$ $(\alpha \leqslant t \leqslant \beta)$, where γ lies in D. Since γ is smooth, $z'(t) = x'(t)+iy'(t)$ exists and is continuous for $\alpha \leqslant t \leqslant \beta$.

Define

$$\int_\gamma f(z)\,dz = \int_\alpha^\beta f(z)\frac{dz}{dt}\,dt \tag{1}$$

By writing $f(z) = u(x, y)+iv(xy)$, we have

$$f(z(t)) = u(x(t), y(t))+iv(x(t), y(t))$$
$$= \underline{u}(t)+i\underline{v}(t),$$

and so equation (1) becomes

$$\int_\gamma f(z)\,dz = \int_\alpha^\beta (\underline{u}(t)+i\underline{v}(t))\,(x'(t)+iy'(t))dt \tag{2}$$

i.e. $$\int_\gamma f(z)\,dz = \int_\alpha^\beta (\underline{u}x'-\underline{v}y')dt+i\int_\alpha^\beta (\underline{u}y'+\underline{v}x')\,dt \tag{3}$$

Equation (3) states that to integrate along a smooth path, we substitute in terms of the parameter t, separate into real and imaginary parts, and calculate two real integrals.

EXAMPLE 1. Integrate z^2 along the smooth path γ given by $z(t) = t+it^2 \qquad (0\leqslant t\leqslant 1)$.

Since $z'(t) = 1+2it$, we have

$$\int_\gamma z^2dz = \int_0^1 (t+it^2)^2\,(1+2it)dt$$
$$= \int_0^1 (t^2-5t^4)dt+i\int_0^1 (4t^3-2t^5)dt$$
$$= [\tfrac{1}{3}t^3-t^5]_0^1+i[t^4-\tfrac{1}{3}t^6]_0^1$$
$$= \tfrac{2}{3}(i-1).$$

EXAMPLE 2. Integrate $1/z$ around the unit circle C given by $z(t) = \cos t+i\sin t \qquad (0\leqslant t\leqslant 2\pi)$.

Since $z'(t) = -\sin t+i\cos t = i(\cos t+i\sin t)$, we have

$$\int_C 1/z\,dz = \int_0^{2\pi} (\cos t+i\sin t)^{-1}\,i(\cos t+i\sin t)\,dt$$

$$= i\int_0^{2\pi} dt$$

$$= 2\pi i.$$

If γ^* is the opposite path to γ, we have

$$\int_{\gamma^*} f(z)\,dz = \int_\alpha^\beta f(z(\alpha+\beta-t))\,\frac{d}{dt}(z(\alpha+\beta-t))\,dt.$$

Put $\alpha+\beta-t = s$, then the integral becomes

$$\int_\beta^\alpha f(z(s))\,\frac{d}{ds}(z(s))\,\frac{ds}{dt}\,\frac{dt}{ds}\,ds$$

$$= \int_\beta^\alpha f(z(s))\,\frac{d}{ds}(z(s))\,ds$$

$$= -\int_\alpha^\beta f(z(s))\,\frac{d}{ds}(z(s))\,ds$$

i.e.
$$\int_{\gamma^*} f(z)\,dz = -\int_\gamma f(z)\,dz. \tag{4}$$

Now suppose γ is a contour. Then γ consists of a finite number of smooth paths $\gamma_1, \ldots\ldots, \gamma_n$ and we define

$$\int_\gamma f(z)\,dz = \int_{\gamma_1} f(z)\,dz + \ldots\ldots + \int_{\gamma_n} f(z)\,dz \tag{5}$$

EXAMPLE 3. Integrate z^2 along the contour γ given by

$$z(t) = \begin{cases} t & (0 \leqslant t \leqslant 1) \\ 1+i(t-1) & (1 \leqslant t \leqslant 2) \end{cases}$$

$$\int_\gamma z^2 dz = \int_0^1 t^2.1dt + \int_1^2 (1+i(t-1))^2.\,idt$$

$$= [\tfrac{1}{3}t^3]_0^1 + \int_1^2 (2-2t)\,dt + i\int_1^2 (2t-t^2)\,dt$$

$$= \tfrac{1}{3} + [2t-t^2]_1^2 + i[t^2-\tfrac{1}{3}t^3]_1^2$$

$$= \tfrac{2}{3}(i-1).$$

Using (4), (5) we have the following rules for contour integration:

RULE 1. If γ is composed of two contours γ_1, γ_2,

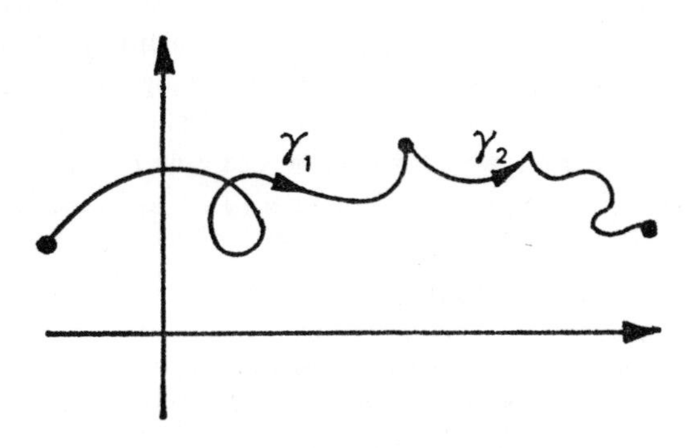

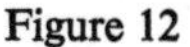

Figure 12

then $\int_\gamma f(z)\,dz = \int_{\gamma_1} f(z)\,dz + \int_{\gamma_2} f(z)\,dz$.

RULE 2. If γ^* is the opposite contour to γ, then

$$\int_{\gamma^*} f(z)\,dz = -\int_\gamma f(z)\,dz.$$

The value of a contour integral is unchanged when we change the parameter $t = h(u)$ $(a \leqslant u \leqslant b)$, where $h(a) = \alpha$, $h(b) = \beta$, h is (strictly) monotonic increasing and $h'(u)$ is continuous for $a \leqslant u \leqslant$ b. We then have

$$\int_\gamma f(z)\,dz = \int_\alpha^\beta f(z)\frac{dz}{dt}\,dt = \int_a^b f(z)\frac{dz}{dt}\frac{dt}{du}\,du = \int_a^b f(z)\frac{dz}{du}\,du.$$

This is analogous to the real case†, and may be used to simplify the calculation.

We could consider two contours to be equivalent for the

† W. Ledermann, *Integral Calculus*, p. 12

purposes of integration if they are related to each other by a change in parameter as above. Two equivalent contours have the same track and are traversed in the same direction as the parameter increases. A particularly simple change in parameter is given by $h(u) = mu+c$ where m, c are real constants and $m>0$. This is called a *linear* change and evidently satisfies the required conditions. By a suitable linear change in parameter we could replace a contour by an equivalent one defined on any parametric interval we please. For example $t = (\beta-\alpha)u+\alpha$ changes the parametric interval from $\alpha \leqslant t \leqslant \beta$ to $0 \leqslant u \leqslant 1$.

3. The Fundamental Theorem

If f is the derivative of an analytic function, then the calculation of $\int_\gamma f(z)\, dz$ is very simple indeed, for we have (analogous to the real case):

THE FUNDAMENTAL THEOREM OF CONTOUR INTEGRATION. Suppose that f is a continuous function defined in the domain D. If f is the derivative of an analytic function F in D, and γ is a contour in D starting at z_1 and ending at z_2, then

$$\int_\gamma f(z)\, dz = F(z_2)-F(z_1).$$

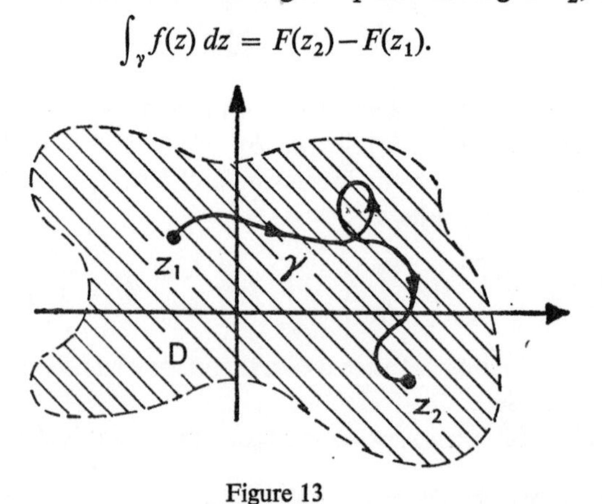

Figure 13

Proof. Write $f(z) = u(x, y)+iv(x, y)$, $F(z) = U(x, y)+iV(x, y)$ and suppose that γ is given by $z(t) = x(t)+iy(t)$ $(\alpha \leqslant t \leqslant \beta)$, where $z(\alpha) = z_1$, $z(\beta) = z_2$.

Since $f = F'$, using the Cauchy-Riemann equations for F, we have

$$F' = u+iv = \frac{\partial U}{\partial x}+i\frac{\partial V}{\partial x} = \frac{\partial V}{\partial y}-i\frac{\partial U}{\partial y}$$

and so $u = \dfrac{\partial U}{\partial x} = \dfrac{\partial V}{\partial y}$, $\quad v = \dfrac{\partial V}{\partial x} = -\dfrac{\partial U}{\partial y}$.

Thus

$$\int_\gamma f(z)\,dz = \int_\alpha^\beta (u+iv)(x'+iy')\,dt$$

$$= \int_\alpha^\beta \left(\frac{\partial U}{\partial x}x'+\frac{\partial U}{\partial y}y'\right)dt+i\int_\alpha^\beta \left(\frac{\partial V}{\partial x}x'+\frac{\partial V}{\partial y}y'\right)dt.$$

But we have† $\dfrac{dU}{dt} = \dfrac{\partial U}{\partial x}\dfrac{dx}{dt}+\dfrac{\partial U}{\partial x}\dfrac{dy}{dt}$

and so $$\int_\gamma f(z)\,dz = \int_\alpha^\beta \frac{dU}{dt}\,dt+i\int_\alpha^\beta \frac{dV}{dt}\,dt$$

$$= F(z(\beta))-F(z(\alpha))$$

$$= F(z_2)-F(z_1), \text{ as required.}$$

If $f = F'$ in D, then F is called a *primitive* of f in D. A primitive is unique up to an additive constant, because if $f = F_1'$, $f = F_2'$ in D, then $F_1'-F_2' = 0$ and so F_1-F_2 is constant in D.

Obviously the simplest way to integrate is to look for a primitive. For example $z^2 = \dfrac{d}{dz}(\frac{1}{3}z^3)$ and $\frac{1}{3}z^3$ is a primitive for z^2. Thus if γ is any contour joining 0 to $1+i$, we have

† P. Hilton, *Partial Derivatives*, pp. 12, 13.

$$\int_\gamma z^2\, dz = \tfrac{1}{3}(1+i)^3 = \tfrac{2}{3}(1-i).$$

This integral has already been calculated for certain contours (examples 1, 3).

More generally, if n is an integer, $n \neq -1$, then $z^n = \dfrac{d}{dz}\left(\dfrac{z^{n+1}}{n+1}\right)$. This holds for all $n \geqslant 0$ and for $z \neq 0$ if $n \leqslant -2$. Hence if γ is a contour not passing through the origin which starts at z_1 and ends at z_2, then

$$\int_\gamma z^n\, dz = \frac{z_2^{n+1}}{n+1} - \frac{z_1^{n+1}}{n+1} \qquad (n \neq -1).$$

In particular, if γ is a closed contour, we have $z_1 = z_2$ and

$$\int_\gamma z^n\, dz = 0 \qquad (n \neq -1).$$

This illustrates the following consequences of the fundamental theorem:

COROLLARY 3.1. If f is the continuous derivative of an analytic function, then for any contour γ in the domain of definition of f, $\int_\gamma f(z)\, dz$ depends only on the endpoints and not on the particular contour.

COROLLARY 3.2. If f is the continuous derivative of an analytic function, then for any *closed* contour γ in the domain of definition, $\int_\gamma f(z) = 0$.

Proof of 3.2. $\int_\gamma f(z)\, dz = F(z_2) - F(z_1) = 0$, since $z_1 = z_2$.

Warning. Not every continuous function has a primitive. For such functions the fundamental theorem does not apply and the only way to evaluate the integral is by direct calculation from the basic definition. For such functions the integral *does* depend on the path, and the integral round a closed curve may not be zero.

If F is analytic in a domain D and $f = F'$, we will prove later (page 56) that f is also analytic. This remarkable theorem shows that if f is not analytic, then it cannot have a primitive F.

Even if f is analytic in D, it need not have a primitive defined throughout the whole of D. For example $f(z) = 1/z$ in the domain D consisting of all points except the origin. If f had a primitive in D, then $\int_\gamma 1/z\,dz = 0$ for any closed contour γ in D. But for the unit circle C given by $z(t) = \cos t + i \sin t\,(0 \leqslant t \leqslant 2\pi)$, we know (example 2) that $\int_C 1/z\,dz = 2\pi i \neq 0$. This shows that no primitive can exist.

We recall that $1/z = \dfrac{d}{dz}(\operatorname{Log} z)$ in the cut-plane, but this does not lead to a contradiction since the unit circle crosses the negative real axis and so does not lie completely in the cut-plane.

Notice the striking difference between z^n (where n is an integer, $n \neq -1$) and z^{-1}, in that

$$\int_C z^n\,dz = \begin{cases} 2\pi i \text{ if } n = -1, \\ 0 \text{ otherwise.} \end{cases}$$

This result is responsible for much of the theory in Chapters II, III of Functions of a Complex Variable II.

We conclude this section with an inequality which will prove useful later:

THEOREM 3.3. If γ is a contour of length L and $|f(z)| \leqslant M$ on the track of γ, then

$$\left|\int_\gamma f(z)\,dz\right| \leqslant ML.$$

Proof. If γ is given by $z(t) = x(t) + iy(t)$ $(\alpha \leqslant t \leqslant \beta)$, then we recall† that the length of γ is given by

$$L = \int_\alpha^\beta \{(x'(t))^2 + (y'(t))^2\}^{\frac{1}{2}}\,dt = \int_\alpha^\beta |z'(t)|\,dt.$$

† W. Ledermann, *Multiple Integrals*, p. 4.

Assuming the inequality

$$\left|\int_{\alpha}^{\beta} g(t)\,dt\right| \leqslant \int_{\alpha}^{\beta} |g(t)|\,dt \tag{6}$$

for a complex function g of a real variable t, then

$$\begin{aligned}\left|\int_{\gamma} f(z)\,dz\right| &= \left|\int_{\alpha}^{\beta} f(z(t))\,z'(t)\,dt\right| \\ &\leqslant \int_{\alpha}^{\beta} |f(z(t))\,z'(t)|\,dt \\ &\leqslant \int_{\alpha}^{\beta} M|z'(t)|\,dt \\ &= ML\end{aligned}$$

To verify (6), we use a simple trick.

Let $\int_{\alpha}^{\beta} g(t)\,dt = Re^{i\Theta}$ where R, Θ are real, $R \geqslant 0$. Then

$$R = \int_{\alpha}^{\beta} e^{-i\Theta} g(t)dt = \int_{\alpha}^{\beta} U(t)dt + i\int_{\alpha}^{\beta} V(t)\,dt$$

where

$$e^{-i\Theta} g(t) = U(t) + iV(t).$$

Since R is real, $\int_{\alpha}^{\beta} V(t)\,dt = 0$. But now, by the real case†, since $U(t) \leqslant |g(t)|$ we have

$$R = \int_{\alpha}^{\beta} U(t)\,dt \leqslant \int_{\alpha}^{\beta} |g(t)|\,dt, \text{ as required.}$$

REMARK. An upper bound M for $|f(z)|$ on the track of γ can always be found. We have not developed the technique for a neat proof, but we sketch an outline of a proof as follows:

The function $m(t) = |f(z(t))|$ is a continuous real-valued function of t for $\alpha \leqslant t \leqslant \beta$. By continuity at α, $m(t)$ must be bounded for $\alpha \leqslant t \leqslant \alpha + \varepsilon$ where $\varepsilon > 0$. Let x_0 be the upper bound of all points x in $\alpha \leqslant x \leqslant \beta$ such that $m(t)$ is bounded in $\alpha \leqslant t \leqslant x$. By continuity at x_0, $m(t)$ must be bounded in $\alpha \leqslant t \leqslant x_0$. By hypothesis we have $x_0 \leqslant \beta$. We cannot have $x_0 < \beta$ because the continuity of $m(t)$ would imply that $m(t)$ is

† W. Ledermann, *Integral Calculus*, p. 6.

bounded in a neighbourhood of x_0, i.e. for $|t-x| \leqslant \delta$ where $\delta > 0$. Thus we would have $m(t)$ bounded for $\alpha \leqslant t \leqslant x_0 + \delta$ (by the larger of the bounds in $\alpha \leqslant t \leqslant x_0$, $x_0 \leqslant t \leqslant x_0 + \delta$). This contradicts the definition of x_0. Hence $x_0 = \beta$ and since $m(t)$ is bounded for $\alpha \leqslant t \leqslant x_0$, we have the desired result.

4. Cauchy's Theorem

In the last section we were discussing conditions under which $\int_\gamma f(z)\,dz = 0$ for a closed contour γ. If f is assumed analytic in a domain containing γ. this need not be true. For example we have seen that $\int_C 1/z\,dz \neq 0$ where C is the unit circle $z(t) = \cos t + i \sin t$ $(0 \leqslant t \leqslant 2\pi)$. The significant factor here is that $1/z$ is not analytic everywhere *inside* C. (It is not defined at the origin.)

CAUCHY'S THEOREM. If f is analytic in a domain D and γ is a closed Jordan contour in D whose interior also lies in D, then

$$\int_\gamma f(z)\,dz = 0.$$

It is not possible in this text to give a complete proof of this

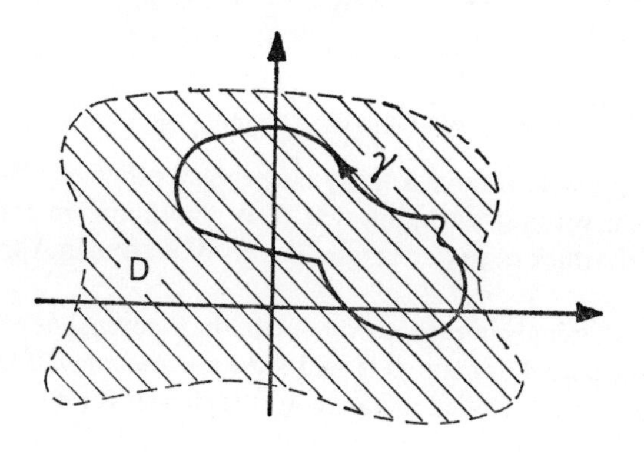

Figure 14

very deep result, but an outline will be given at the end of this section. In the original proof, Cauchy himself needed to assume that not only was f analytic (i.e. f' exists throughout D) but that f' was continuous. He gave several proofs, one of which used the Cauchy-Riemann equations.

$$\int_\gamma f(z)dz = \int_\gamma (u+iv)(dx+idy)$$
$$= \int_\gamma (u\,dx - v\,dy) + i\int_\gamma (v\,dx + u\,dy).$$

Let A be the set of points inside and on the track of γ. Then Green's Theorem† states that under suitable conditions, including the continuity of $P(x, y)$, $Q(x, y)$, $\frac{\partial Q}{\partial x}$, $\frac{\partial P}{\partial y}$, we have

$$\int_\gamma (P\,dx + Q\,dy) = \iint_A \left(\frac{\partial Q}{\partial x} - \frac{\partial P}{\partial y}\right) dx\,dy.$$

If we assume f' is continuous, this implies the continuity of u, v, $\frac{\partial u}{\partial x}$, $\frac{\partial u}{\partial y}$, $\frac{\partial v}{\partial x}$, $\frac{\partial v}{\partial y}$. Hence

$$\int_\gamma f(z)\,dz = -\iint_A \left(\frac{\partial v}{\partial x} + \frac{\partial u}{\partial y}\right) dx\,dy + i\iint_A \left(\frac{\partial u}{\partial x} - \frac{\partial v}{\partial y}\right) dx\,dy$$

$= 0$ by the Cauchy-Riemann equations.

It is possible to give a fairly elementary proof of Cauchy's Theorem without assuming that f' is continuous in the case where the track of γ is a triangle. The proof is given in Appendix II.

This seemingly innocuous version of the theorem has a strong consequence. A domain D is said to be a *star-domain* if there is a point z_0 in D (called a star-centre) such that for every other

† W. Ledermann, *Multiple Integrals*, p. 38.

point z in D, the whole straight line segment joining z_0 to z lies in D. Examples of star-domains are drawn in figure 15:

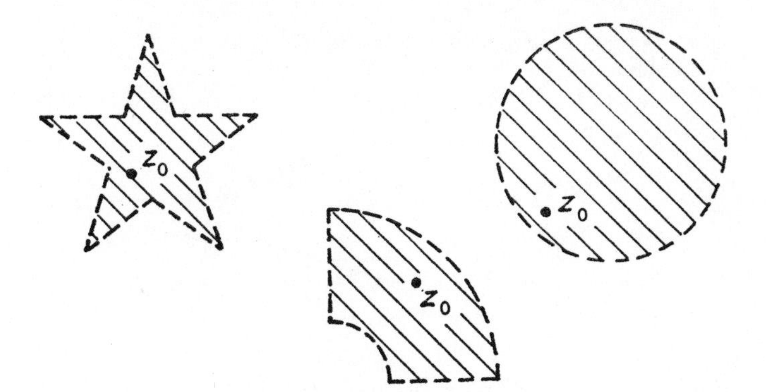

Figure 15

Inside a star-domain an analytic function always has a primitive:

PROPOSITION 4.1. If f is an analytic function defined in a star-domain D, then we may construct an analytic function F defined in D such that $F' = f$.

Proof. Denote by $[z_1, z_2]$ the contour $z(t) = z_1(1-t)+z_2t$ $(0 \leqslant t \leqslant 1)$ which describes the straight line joining z_1 to z_2. If z_0 is the star-centre of D and z_1 is in D, then $[z_0, z_1]$ lies in D and we define

$$F(z_1) = \int_{[z_0, z_1]} f(z)\, dz$$

We will prove $F' = f$ and so F is a primitive for f.

Since D is open, there is an $\varepsilon_1 > 0$ such that for $|h| < \varepsilon_1$ we have $z_1 + h$ is in D, and evidently the line $[z_1, z_1 + h]$ lies in D. By Cauchy's Theorem for a Triangle (Appendix II), we have

$$\int_{[z_0, z_1]} f(z)\, dz + \int_{[z_1, z_1+h]} f(z)\, dz + \int_{[z_1+h, z_0]} f(z)\, dz = 0.$$

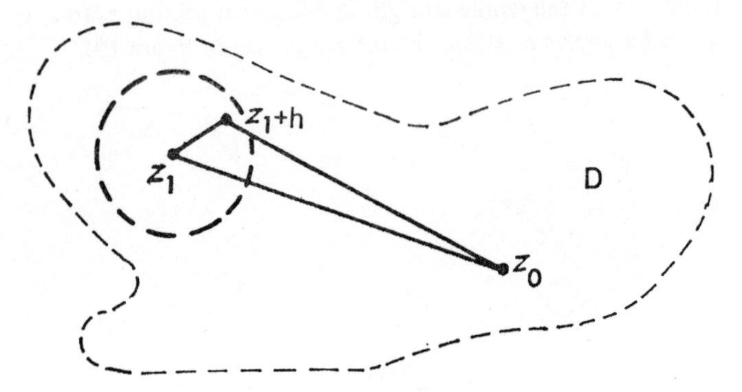

Figure 16

and so $F(z_1+h)-F(z_1) = \int_{[z_0,z_1+h]} f(z)\,dz - \int_{[z_0,z_1]} f(z)\,dz$

$$= \int_{[z_1,z_1+h]} f(z)\,dz.$$

Keeping z_1 constant, we have $\int_{[z_1,z_1+h]} f(z_1)\,dz = f(z_1)h$ and this gives

$$\frac{F(z_1+h)-F(z_1)}{h} - f(z_1) = \int_{[z_1,z_1+h]} \frac{\{f(z)-f(z_1)\}}{h}\,dz \quad (h \neq 0).$$

Since f is analytic in D, it is certainly continuous at z_1 and so given $\varepsilon>0$, we have $|f(z)-f(z_1)|<\varepsilon$ for z in a neighbourhood of z_1. Also the length of $[z_1, z_1+h]$ is $|h|$ and so for sufficiently small h we have

$$\left|\frac{F(z_1+h)-F(z_1)}{h} - f(z_1)\right| \leqslant \frac{\varepsilon}{|h|}\cdot|h| = \varepsilon.$$

Since ε is arbitrary, this implies

$$\lim_{h\to 0} \frac{F(z_1+h)-F(z_1)}{h} = f(z_1)$$

i.e. $F'(z_1) = f(z_1)$.

Since z_1 is an arbitrary point in D, this completes the proof.

As immediate consequences of this proposition and corollaries 3.1, 3.2 of the fundamental theorem of contour integration, we have:

COROLLARY 4.2. If f is an analytic function defined in a star-domain D, then for any contour γ in D, $\int_\gamma f(z)\,dz$ depends only on the end-points of γ.

COROLLARY 4.3. If f is analytic in a star-domain D and γ is any closed contour in D, then $\int_\gamma f(z)\,dz = 0$.

We may use corollary 4.2 to sketch a proof of Cauchy's Theorem. First note that an open disc given by $|z-z_0|<r$ is a star-domain. If γ is a contour in an arbitrary domain D, it is always possible to subdivide γ into a finite number of sub-contours $\gamma_1, \ldots, \gamma_n$ where each γ_r lies in an open disc D_r which itself is contained in D. (The proof is omitted.) Let z_{r-1}, z_r be the initial and final points of γ_r.

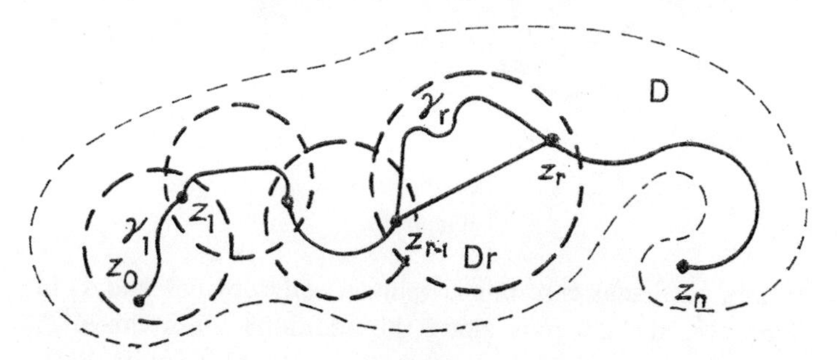

Figure 17

Since D_r is a star-domain, by corollary 4.2,

$$\int_{\gamma_r} f(z)\,dz = \int_{[z_{r-1},z_r]} f(z)\,dz.$$

If P is the polygon with sides $[z_0, z_1], [z_1, z_2], \ldots, [z_{n-1}, z_n]$, then

$$\int_\gamma f(z)\,dz = \sum_{r=1}^{n} \int_{\gamma_r} f(z)\,dz = \int_P f(z)\,dz.$$

Now suppose γ is a *closed* Jordan contour in D whose interior also lies in D. To show $\int_\gamma f(z)\,dz = 0$, it is sufficient to prove $\int_P f(z)\,dz = 0$ for the closed polygon P. To do this, we may draw in extra lines joining vertices of P, making triangular contours $\Delta_1, \ldots, \Delta_m$ such that

(i) The track and interior of each Δ_r lies in D,

(ii) $\int_P f(z)\,dz = \sum_{r=1}^{m} \int_{\Delta_r} f(z)\,dz.$

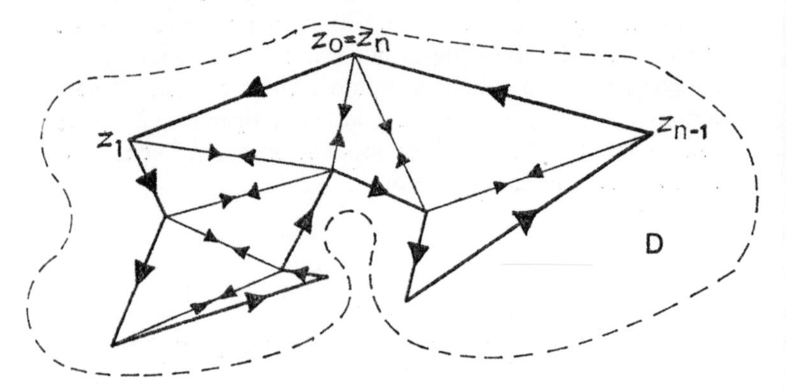

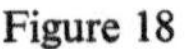
Figure 18

In any particular case this is geometrically obvious and as in figure 18, the integrals along the additional lines cancel in opposite pairs. Note, however, that it is difficult to write down a general rule as to how this is done. Assuming its validity, by (i) and Cauchy's Theorem for a triangle we have $\int_{\Delta_r} f(z)\,dz = 0$ and by (ii) we have $\int_P f(z)\,dz = 0$. Hence $\int_\gamma f(z)\,dz = 0$.

This concludes our discussion on Cauchy's Theorem.

EXERCISES ON CHAPTER TWO

In exercises 1–8 integrate the given function along the contour $z(t) = 1-t+it^2$ $(0 \leqslant t \leqslant 1)$. (Use the Fundamental Theorem of Contour Integration wherever possible.)

1. $\mathscr{R}z$ 2. $1/z^3$ 3. $(4z^3+z^4)e^z$ 4. $1/z$ 5. z^2 6. $\bar{z}$
7. $\sin^2 z$ 8. $z^\alpha = e^{\alpha \operatorname{Log} z}$ $(\alpha \neq -1)$.

In exercises 9–11, integrate the given function around the unit circle $z(t) = \cos t + i \sin t$ $(0 \leqslant t \leqslant 2\pi)$.

9. $1/z^2$ 10. $|z|$ 11. $\bar{z}$.

12. If $f(z) = c_0 + c_1 z + \ldots\ldots + c_n z^n + \ldots\ldots$ for $|z| < R$, prove $F(z) = c_0 z + \dfrac{c_1 z^2}{2} + \ldots + \dfrac{c_n z^{n+1}}{n+1} + \ldots$ is absolutely convergent for $|z| < R$.

Use the result of Appendix I to show $F'(z) = f(z)$ for $|z| < R$. If γ is a contour in $|z| < R$, starting at the origin and finishing at z_0, show

$$\int_\gamma f(z)\, dz = c_0 z_0 + \frac{c_1 z_0^2}{2} + \ldots + \frac{c_n z_0^{n+1}}{n+1} + \ldots$$

(This states a power series may be integrated term by term inside the circle of convergence.)

CHAPTER THREE

Taylor's Series

1. Cauchy's Integral Formula

THEOREM 1.1. Suppose that f is an analytic function defined in a domain D. Let γ be a closed Jordan contour in D whose interior lies completely in D. If γ is described anti-clockwise (as the parameter increases) and z_0 is a point inside γ, then

$$f(z_0) = \frac{1}{2\pi i}\int_\gamma \frac{f(z)}{z-z_0}\,dz.$$

This is Cauchy's Integral Formula.

Proof. Let C_ε be the circle centre z_0, radius ε in the standard parametrization $z(t) = z_0+\varepsilon e^{it}$ $(0\leqslant t\leqslant 2\pi)$, where ε is small so that the track of C_ε is inside that of γ.

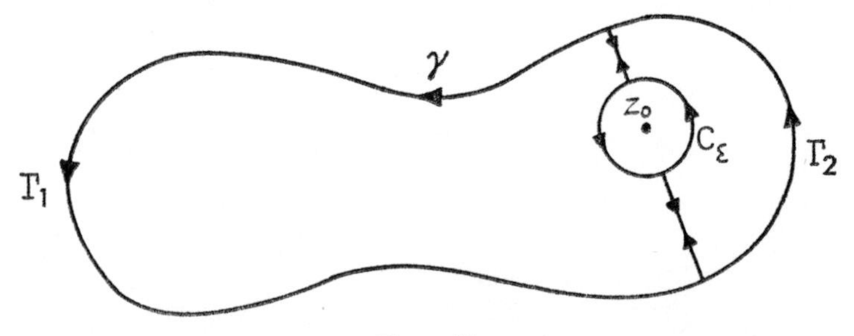

Figure 19

Make two cross-cuts from the track of C_ε to that of γ and parametrize them, making two Jordan contours Γ_1, Γ_2 where Γ_1, Γ_2 each traverse part of γ anti-clockwise, across a cut, round part of C_ε clockwise (i.e. the opposite sense to C_ε) and across the other cut as in figure 19. (We are relying on geometric intuition for this construction.)

The function $F(z) = \dfrac{f(z)}{z-z_0}$ is analytic inside and on Γ_r for $r = 1, 2$ and by Cauchy's Theorem

$$\int_{\Gamma_r} F(z)\, dz = 0 \qquad r = 1, 2.$$

Adding these two integrals, the contributions due to the cross-cuts cancel and we have

$$\int_{\gamma} F(z)\, dz - \int_{C_\varepsilon} F(z)\, dz = 0. \tag{1}$$

Now

$$\int_{C_\varepsilon} F(z)\, dz = \int_{C_\varepsilon} \frac{f(z_0)}{z-z_0}\, dz + \int_{C_\varepsilon} \frac{f(z)-f(z_0)}{z-z_0}\, dz. \tag{2}$$

Since $\lim\limits_{z\to z_0} \dfrac{f(z)-f(z_0)}{z-z_0} = f'(z_0)$, for z near z_0 we must have

$$\left|\frac{f(z)-f(z_0)}{z-z_0}\right| \leqslant M \text{ and so } \left|\int_{C_\varepsilon} \frac{f(z)-f(z_0)}{z-z_0}\, dz\right| \leqslant M.2\pi\varepsilon$$

using theorem 3.3 of Chapter II. As $\varepsilon \to 0$, the contribution of this integral tends to zero.
Also

$$\int_{C_\varepsilon} \frac{f(z_0)}{z-z_0}\, dz = f(z_0) \int_0^{2\pi} \frac{1}{e^{it}}\, ie^{it}\, dt = f(z_0).2\pi i.$$

Substituting into (1), and letting $\varepsilon \to 0$, we find

$$\int_{\gamma} \frac{f(z)}{z-z_0}\, dz = f(z_0).2\pi i,$$

which completes the proof.

Note that this remarkable theorem shows that the values of an analytic function at all points inside a closed Jordan contour are uniquely determined by the values of the function on that contour.

2. Taylor's Series

In this section we use Cauchy's integral formula to express an analytic function as a power series in a neighbourhood of a point.

LEMMA 2.1. If f is analytic in the open disc given by $|z-z_0|<R$, then $f(z_0+h) = a_0+a_1h+\ldots+a_nh^n+\ldots$ for $|h|<R$. If C is a circle centre z_0, radius r where $0<r<R$, given by $z(t) = z_0+re^{it}$ $(0\leqslant t\leqslant 2\pi)$ then $a_n = \dfrac{1}{2\pi i}\displaystyle\int_C \frac{f(z)}{(z-z_0)^{n+1}}dz.$

Proof. Fix h where $|h|<R$ and initially restrict r to $|h|<r<R$. By Cauchy's integral formula

$$f(z_0+h) = \frac{1}{2\pi i}\int_C \frac{f(z)}{z-z_0-h}\,dz.$$

But $1/(z-z_0-h) = \dfrac{1}{z-z_0}\left(1-\dfrac{h}{z-z_0}\right)^{-1} = \dfrac{1}{z-z_0}\,(1-\mathbf{w})^{-1}$

where $\mathbf{w} = \dfrac{h}{z-z_0}$. Since $1+\mathbf{w}+\ldots+\mathbf{w}^{n-1} = \dfrac{1-\mathbf{w}^n}{1-\mathbf{w}}$, we have

$$(1-\mathbf{w})^{-1} = 1+\mathbf{w}+\ldots+\mathbf{w}^{n-1}+\frac{\mathbf{w}^n}{1-\mathbf{w}}.$$

Substituting into the integral formula for $f(z_0+h)$ and simplifying, we obtain

$$f(z_0+h) = \frac{1}{2\pi i}\int_C f(z)\left\{\frac{1}{z-z_0}+\frac{h}{(z-z_0)^2}+\ldots+\frac{h^{n-1}}{(z-z_0)^n}\right.$$

$$\left.+\frac{h^n}{(z-z_0)^n(z-z_0-h)}\right\}dz = a_0+a_1h+\ldots+a_{n-1}h^{n-1}+A_n$$

where

$$a_m = \frac{1}{2\pi i}\int_C \frac{f(z)}{(z-z_0)^{m+1}}\,dz$$

and

$$A_n = \frac{1}{2\pi i}\int_C \frac{f(z)h^n}{(z-z_0)^n(z-z_0-h)}\,dz.$$

For z on the track of C we have $|f(z)| \leqslant M$ where M is some real constant. Moreover for z on the track of C, $|z-z_0| = r$ and $|z-z_0-h| \geqslant ||z-z_0|-|h|| = r-|h|$, hence

$$|A_n| \leqslant \frac{1}{2\pi}\frac{M|h|^n}{r^n(r-|h|)}2\pi r = \frac{Mr}{(r-|h|)}\left(\frac{|h|}{r}\right)^n.$$

Since $|h| < r$, we have $A_n \to 0$ as $n \to \infty$ and so the infinite series $\sum a_n h^n$ converges to the sum $f(z_0+h)$.

Note that we have only proved the expression

$$a_n = \frac{1}{2\pi i}\int_C \frac{f(z)}{(z-z_0)^{n+1}}\,dz$$

for $|h| < r < R$, but since the integral is independent of h, no matter how small, the expression must be true for any r satisfying $0 < r < R$.

If we write $z = z_0+h$ then we have

$$f(z) = a_0+a_1(z-z_0)+\ \dots\ +a_n(z-z_0)^n+\ \dots \text{ for } |z-z_0| < R.$$

But a power series is differentiable as many times as we please inside its circle of convergence (proposition 6.1, Chapter I) and $a_n = \dfrac{f^{(n)}(z_0)}{n!}$.

Now suppose that f is defined on an arbitrary domain D. If z_0 is in D then, by definition, so is an ε-neighbourhood given by $|z-z_0| < \varepsilon$. If R is the largest such ε (possible infinite), then using lemma 2.1 inside $|z-z_0| < R$ we obtain:

TAYLOR'S THEOREM. If f is analytic in a domain D, then

f is differentiable as many times as we please throughout D. If z_0 is in D, then

$$f(z) = f(z_0)+f'(z_0)(z-z_0)+ \dots + \\ + \frac{f^{(n)}(z_0)}{n!}(z-z_0)^n + \dots \qquad |z-z_0|<R,$$

where $|z-z_0|<R$ is the largest open disc centre z_0 contained in D.

Substituting $z = z_0+h$, the power series may also be written as:

$$f(z_0+h) = f(z_0)+f'(z_0)h+ \dots +\frac{f^{(n)}(z_0)}{n!}h^n+ \dots \qquad |h|<R.$$

REMARK. The importance of this phenomenal result cannot be over-emphasized. We need only assume a complex function is differentiable once throughout its domain of definition and then it is infinitely differentiable. This contrasts strongly with the real case, where a function may be differentiable once but not twice (refer back to page 30 for an example).

EXAMPLE 1. $f(z) = \text{Log } z$ is analytic in the cut-plane. The largest open disc centre $z_0 = 1$ in the cut-plane is $|z-1|<1$. Since $f^{(n)}(z_0) = \dfrac{(-1)^n(n-1)!}{z_0^n} = (-1)^n(n-1)!$, we have:

$$\text{Log}\,(1+h) = h-\frac{h^2}{2}+ \dots +\frac{(-1)^n h^n}{n}+ \dots \qquad |h|<1.$$

EXAMPLE 2. $f(z) = 1/z$ is analytic for $z\neq 0$. If $z_0\neq 0$, then

$$1/(z_0+h) = \frac{1}{z_0}\left(1+\frac{h}{z_0}\right)^{-1}$$

$$= \frac{1}{z_0}-\frac{h}{z_0^2}+\frac{h^2}{z_0^3}- \dots +\frac{(-1)^n h^n}{z_0^{n+1}}+ \dots \qquad \left|\frac{h}{z_0}\right|<1.$$

If $\left|\frac{h}{z_0}\right| < 1$, then $|h| < |z_0|$, which states that z_0+h lies in the circle centre z_0, radius $|z_0|$. This is the largest circle centre z_0 which does not include the origin (where $1/z$ is not defined). Note that the coefficient of h^n in the power series is

$$\frac{(-1)^n}{z_0^{n+1}} = \frac{1}{n!} f^{(n)}(z_0).$$

In the notation of lemma 2.1,

$$f^{(n)}(z_0) = n!a_n = \frac{n!}{2\pi i}\int_C \frac{f(z)}{(z-z_0)^{n+1}}\, dz,$$

where C is a circle centre z_0, lying in the domain of definition of f. This is Cauchy's Formula for the nth derivative of f.

If $|f(z)| \leqslant M$ on the circle C centre z_0, radius r, then

$$|f^{(n)}(z_0)| = \frac{n!}{2\pi}\left|\int_C \frac{f(z)}{(z-z_0)^{n+1}}\, dz\right|$$
$$\leqslant \frac{n!}{2\pi}\frac{M}{r^{n+1}}.2\pi r$$

and so

$$|f^{(n)}(z_0)| \leqslant \frac{Mn!}{r^n}.$$

This result is called *Cauchy's Inequality.* Using it we may prove:

LIOUVILLE'S THEOREM. If f is analytic throughout the whole plane and $|f(z)| \leqslant M$ for all z, then f is a constant function.

Proof. $|f'(z_0)| \leqslant \frac{M}{r}$ for *any* r, since f is analytic throughout the whole plane. Let $r \to \infty$ and we have $f'(z_0) = 0$. This is true for any z_0 and so $f'(z) = 0$ for all z. This implies that f is constant.

3. Zeros and the Identity Theorem

If $f(z_0) = 0$, we say that z_0 is a *zero of f*. We may write an analytic function f as a power series in a neighbourhood of z_0,

$$f(z) = a_0+a_1(z-z_0)+ \dots +a_n(z-z_0)^n+ \dots \quad |z-z_0|<R.$$

Either $a_n = 0$ for all n, in which case $f(z) = 0$ for $|z-z_0|<R$, or we have $a_0 = a_1 = \dots = a_{m-1} = 0$ and $a_m \neq 0$. In the latter case we say that z_0 is a *zero of order m*. Note that since $a_n = f^{(n)}(z_0)/n!$, a zero of order m satisfies $f(z_0) = 0, f'(z_0) = 0, \dots, f^{(m-1)}(z_0) = 0$, but $f^{(m)}(z_0) \neq 0$.

We may show that a zero of order m is isolated. By this we mean that there is an ε-neighbourhood of z_0 in which z_0 is the *only* zero of f, i.e. $f(z) \neq 0$ for $0<|z-z_0|<\varepsilon$.

To see this we write

$$\begin{aligned} f(z) &= (z-z_0)^m \{a_m+a_{m+1}(z-z_0)+ \dots\} \text{ for } |z-z_0|<R \\ &= (z-z_0)^m \Phi(z) \end{aligned}$$

where the power series $\Phi(z) = a_m+a_{m+1}(z-z_0)+ \dots$ is convergent for $|z-z_0|<R$. Since Φ is analytic for $|z-z_0|<R$, it is certainly continuous at z_0 and so $\Phi(z) \to a_m \neq 0$ as $z \to z_0$. Hence $\Phi(z)$ is non-zero in some ε-neighbourhood of z_0. But $(z-z_0)^m$ is zero only at z_0 and so $f(z) \neq 0$ for $0<|z-z_0|<\varepsilon$.

Suppose that $z_1, z_2, \dots, z_n, \dots$ is a sequence of distinct† zeros of f which tends to a point z_0. If f is defined at z_0, then by continuity we must have $f(z_0) = 0$. Since z_0 is a limit of zeros, it is not isolated, and as we have seen above we must have f identically zero inside some circle centre z_0.

Note. This argument depends on the fact that f is analytic in a neighbourhood of z_0, in particular it breaks down if f is

† We consider the zeros to be distinct to avoid the trivial case that all but a finite number of the zeros coincide at z_0.

not defined or not analytic at z_0. For example the function f given by $f(z) = \sin(1/z)$ $(z \neq 0)$ is analytic everywhere except at the origin. It has a sequence of zeros given by $z_n = 1/n\pi$ $(n \geqslant 1)$ which tends to the origin, but evidently the function is not identically zero inside any circle with the origin as centre.

THEOREM 3.1. Suppose that $z_1, z_2, \ldots, z_n, \ldots$ is a sequence of distinct zeros of an analytic function f defined in a domain D and that the limit of this sequence, z_0, lies in D, then f is identically zero throughout D.

Proof. By continuity $f(z_0) = 0$ and, as we have seen above, f is identically zero inside some circle $|z - z_0| < \varepsilon_0$.

Let w be any other point in D. Since D is a domain, there is a stepwise curve in D joining z_0 to w. We suppose that this curve has length d and let $z(s)$ be the point distance s along it from z_0 so that $z(0) = z_0$ and $z(d) = w$. We intend to show that $f(z) = 0$ all along the curve, in particular $f(w) = 0$.

Since f is identically zero in $|z - z_0| < \varepsilon_0$, then $f(z(s))$ is certainly zero for $0 \leqslant s < \varepsilon_0$. We consider those real numbers s in $0 \leqslant s \leqslant d$ such that $f(z) = 0$ along the curve as far as $z(s)$. Suppose that s^* is the least upper bound of such s. By continuity $f(z(s^*)) = 0$ and $z(s^*)$ is the furthest point along the curve such that $f(z) = 0$ for all z on the curve as far as $z(s^*)$ (marked with a thick line in figure 20).

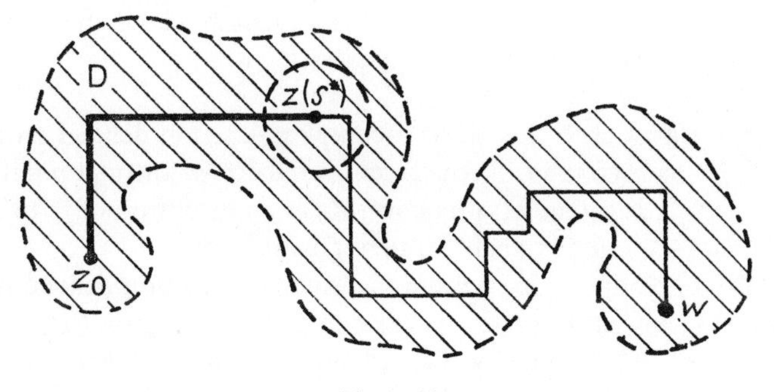

Figure 20

We cannot have $z(s^*) \neq w$. This is because $f(z) = 0$ along the curve up to $z(s^*)$ and so $f(z) = 0$ in a neighbourhood $|z - z(s^*)| < \varepsilon$. This would imply that $f(z) = 0$ for a certain distance along the curve beyond $z(s^*)$, contradicting the definition of s^*. Hence $z(s^*) = w$ and $f(w) = 0$. This completes the proof.

We may immediately deduce:

THE IDENTITY THEOREM. Suppose that f, g are analytic functions defined in the same domain D. Let $z_1, z_2, \ldots, z_n, \ldots$ be a sequence of distinct points in D with limit z_0 also in D, such that $f(z_n) = g(z_n)$ for $n \geqslant 1$, then $f(z) = g(z)$ throughout D.

Proof. Apply theorem 3.1 to $\Phi(z) = f(z) - g(z)$, then Φ is analytic in D and $z_1, z_2, \ldots, z_n, \ldots$ is a sequence of distinct zeros of Φ with limit z_0 in D.

The Identity Theorem has far reaching consequences in the theory of analytic functions.

Suppose that f_0 is a complex function defined on a set S. A complex function f is said to be an *extension* of f_0 if f is defined on a larger set D containing S and $f(z) = f_0(z)$ for all z in S. In general the values of f at points outside S can be assigned quite arbitrarily. For example if S is the real axis and $f(x) = \sin x$ for x real, then defining $f(z) = \sin z$ for z on the real axis and $f(z) = 0$ otherwise, the function f is an extension of f_0. However if we insist that the extension is also *analytic*, then (under a minor restriction on S) the Identity Theorem shows that this extension is *unique*.

THEOREM 3.2. Let f_0 be a complex function defined on a set S which contains a convergent sequence of distinct points $z_1, z_2, \ldots$ together with its limit. If f is an extension of f_0 to a domain D and f is analytic, then f is unique.

Proof. Suppose that g is another analytic function defined in D satisfying $g(z) = f_0(z)$ for all z in S. Then $g(z_n) = f_0(z_n) = f(z_n)$ for $n \geqslant 1$ and by the Identity Theorem, $g(z) = f(z)$ throughout D.

We remark that the notion of extension to an analytic function does guarantee that such an extension exists, only that if it exists then its uniqueness is assured. It also does not give any practical method of constructing an extension and we will find that usually the most successful way is to resort to inspired guesswork.

EXAMPLE. $f_0(z) = 1+z+z^2+\ldots+z^n+\ldots \quad |z|<1.$
The set S of complex numbers satisfying $|z|<1$ certainly contains a convergent sequence of points $\left(\text{e.g. } \frac{1}{2}, \frac{1}{3}, \ldots, \frac{1}{n+1}, \ldots \text{ with limit } 0\right)$ and so an extension to an analytic function in any given domain D is unique. The power series for f_0 is not convergent for $|z|>1$ but the function $f(z) = (1-z)^{-1}$ is analytic for $z\neq 1$ and satisfies $f(z) = f_0(z)$ for $|z|<1$. Hence the analytic function f is the extension of f_0 to the domain consisting of all complex numbers except $z = 1$. Note however that no analytic function exists which is an extension to the whole plane, because $f(z)$ has no finite limit as $z \to 1$ and so we cannot define $f(1)$ in any way to make f analytic there.

The notion of extension by an analytic function is particularly interesting in two cases:

CASE I. S is the real axis (or more generally any subset of the real axis containing a convergent sequence of distinct points together with its limit). Given a real-valued function f_0 defined on S, if there is an extension to a complex analytic function f in some domain containing S then this function is unique. This shows the strong restriction imposed on a complex function by requiring it to be analytic.

For example if $f_0(x) = \sin x$ for all real x, then of course we know that $f(z) = \sin z$ gives an analytic function defined throughout the whole plane and f coincides with f_0 on the real axis. We now know that $f(z) = \sin z$ is the *only* analytic function which satisfies $f(x) = \sin x$ for x real.

CASE II. S is a non-empty open set.

Since S is non-empty it contains a point z_0 and since it is open it includes an ε-neighbourhood of z_0. We can easily select a sequence of distinct points in this neighbourhood which tends to z_0 $\left(\text{e.g. } z_0+\frac{1}{2}\varepsilon,\ z_0+\frac{1}{3}\varepsilon,\ \ldots,\ z_0+\frac{1}{n+1}\varepsilon,\ \ldots\right)$ and so S satisfies the required conditions. As a particular instance we may take S to be the open disc $|z-z_0|<\varepsilon$.

We have seen in section 2 that if f is an analytic function defined in a domain D and z_0 is in D, then f has a Taylor series expansion in a small disc centre z_0. The notion of extension using an analytic function shows that the reverse process is true in that once we know the values in a small disc in D then the values of f throughout D are uniquely determined. Hence in some peculiar way the power series expansion in a small disc contains all the information required to specify the values of the function throughout its domain of definition!

EXERCISES ON CHAPTER THREE

Find the Taylor expansion of the following analytic functions at the origin:

1. $z(1-z)^{-2}$ 2. z^3e^z 3. $(z+1)^3$ 4. $\text{Log}\,(1+z)$ 5. $(1+z)^\alpha$ 6. $(1+z^2)^{-1}$.
7. Suppose that f is analytic throughout the whole plane and satisfies $|f(z)| \leqslant M|z|^n$ for all z. Use Cauchy's inequality to prove that $f^{(n+1)}(z) = 0$ and show that $f(z)$ is polynomial of degree at most n.

Find the extension of the power series in 8–10 to analytic functions in the largest possible domain.

8. $\sum_{n=1}^{\infty} nz^n \quad |z|<1$ 9. $\sum_{n=1}^{\infty} n^2z^n \quad |z|<1$ (Hint: differentiate $\sum nz^n$)
10. $\sum_{n=0}^{\infty} (-1)^n z^{2n} \quad |z|<1$.
11. If $a_n(z) = z^n+(1-z)^n$, by considering $\sum_{n=0}^{N} a_n(z)$, prove that $\sum_{n=0}^{\infty} a_n(z)$ converges if $|z|<1$ and $|1-z|<1$. Draw the domain given by $|z|<1$ and $|1-z|<1$. Find the sum $f(z) = \sum_{n=0}^{\infty} a_n(z)$ in this domain and hence write down the extension of $f(z)$ to an analytic function in the largest possible domain.
12. Suppose that f is analytic in a domain D containing the point $z = 1$ and $f\left(1-\frac{1}{n}\right) = \sum_{n=0}^{\infty} (-1)^n \left(1-\frac{1}{n}\right)^{2n}$ for $n = 1, 2, \ldots$.

    Calculate the following (if they exist):

    $$f(0),\ f(1+i),\ f(i),\ f(2{,}000).$$

Appendix I

THEOREM. If $f(z) = c_0+c_1z+c_2z^2+ \dots +c_nz^n+ \dots$ for $|z|<R$, then $f'(z) = c_1+2c_2z+ \dots +nc_nz^{n-1}+ \dots$ for $|z|<R$.

Proof. First we show that the power series

$$f_1(z) = c_1+2c_2z+ \dots +nc_nz^{n-1}+ \dots$$

is absolutely convergent for $|z|<R$.

Fix z and choose r such that $|z|<r<R$.

By hypothesis $\sum_{n=0}^{\infty} c_nr^n$ converges absolutely and so there is some positive number K such that $|c_nr^n|<K$ for all n.

Let $q = \dfrac{|z|}{r}$, then $0\leqslant q<1$ and $|nc_nz^{n-1}|\leqslant \dfrac{nK|z|^{n-1}}{r^n} = \dfrac{Knq^{n-1}}{r}$.

But $\sum_{n=0}^{\infty} nq^{n-1}$ converges (to $(1-q)^{-2}$), hence by the comparison test, $\sum_{n=0}^{\infty} nc_nz^{n-1}$ converges absolutely.

Now we show $f'(z_0) = f_1(z_0)$ for $|z_0|<R$, i.e.

$$\lim_{z\to z_0}\left\{\frac{f(z)-f(z_0)}{z-z_0}-f_1(z_0)\right\} = 0.$$

As before, choose r such that $|z_0|<r<R$ and since $z\to z_0$, we may also restrict z so that $|z|<r$.

We know $\sum_{n=0}^{\infty} nc_nr^{n-1}$ converges absolutely. Suppose we are given $\varepsilon>0$, then we can find an integer N such that $\sum_{n=N}^{\infty}|nc_nr^{n-1}|$ $<\frac{1}{4}\varepsilon$. Now keep N fixed. We can write

$$\frac{f(z)-f(z_0)}{z-z_0}-f_1(z) = \sum_{n=0}^{\infty} c_n \{z^{n-1}+z_0 z^{n-2}+ \dots +z_0{}^{n-1}-nz_0{}^{n-1}\}$$

We let $\sum_1$ be the sum of the first N terms of this series (i.e. from $n = 0$ to $n = N-1$) and $\sum_2$ the sum of the remaining terms. Then

$$|\textstyle\sum_2| \leqslant \displaystyle\sum_{n=N}^{\infty} |c_n| \{r^{n-1}+r^{n-1}+ \dots +r^{n-1}+nr^{n-1}\}$$
$$= \sum_{n=N}^{\infty} 2n|c_n|r^{n-1} < \tfrac{1}{2}\varepsilon.$$

Also $\sum_1 = \sum_{n=0}^{N} c_n \{z^{n-1}+z_0 z^{n-2}+ \dots +z_0{}^{n-1}-nz_0{}^{n-1}\}$ is a polynomial in z and $\lim_{z\to z_0} \sum_1 = 0$. Hence there is a $\delta>0$ such that $|\sum_1|<\frac{1}{2}\varepsilon$ provided that $|z-z_0|<\delta$. Thus for $|z|<r$ and $|z-z_0|<\delta$ we have

$$\left|\frac{f(z)-f(z_0)}{z-z_0}-f_1(z_0)\right| \leqslant |\textstyle\sum_1|+|\sum_2| < \tfrac{1}{2}\varepsilon+\tfrac{1}{2}\varepsilon = \varepsilon.$$

This means $f'(z_0) = f_1(z_0)$ as required.

Appendix II

CAUCHY'S THEOREM FOR A TRIANGLE

Suppose f is analytic in a domain D and T is a triangular contour whose track and interior lie in D, then $\int_T f(z)\,dz = 0$.

Proof. Suppose $|\int_T f(z)\,dz| = h \geqslant 0$, then by a neat trick we show $h = 0$.

Draw in lines joining the midpoints of the sides of T and parametrize them, giving four triangular contours $T^{(1)}$, $T^{(2)}$, $T^{(3)}$, $T^{(4)}$, such that integrals along the additional lines cancel in pairs because they are taken in opposite directions.

If $I_n = \int_{T^{(n)}} f(z)\,dz$, $n = 1, 2, 3, 4$,
then

$$I_1+I_2+I_3+I_4 = \int_T f(z)\,dz.$$

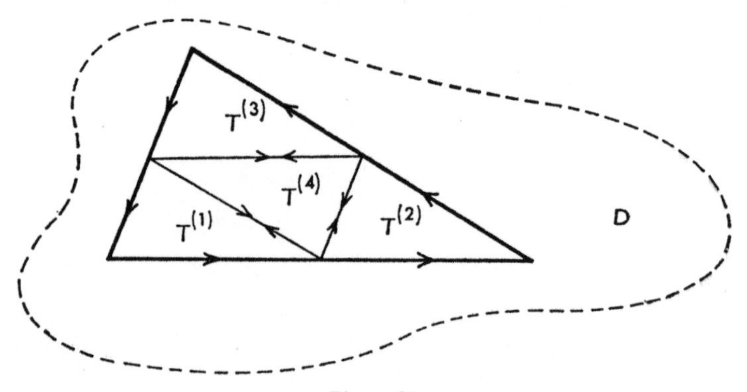

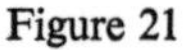

Figure 21

Since $|\int_T f(z)\,dz| = h$, we can choose r such that $|I_r| \geqslant \frac{1}{4}h$. Define $T_1 = T^{(r)}$, then

$$\left|\int_{T_1} f(z)\,dz\right| \geqslant \tfrac{1}{4}h$$

and since T_1 is half the linear size of T, the perimeter length of T_1 is given by

$$L(T_1) = \tfrac{1}{2}L(T).$$

Repeat the process of subdivision with T_1 and so on, obtaining a sequence of triangles $T_1, T_2, \ldots, T_n, \ldots$ where

$$\left|\int_{T_n} f(z)\,dz\right| \geqslant (\tfrac{1}{4})^n h \tag{1}$$

and

$$L(T_n) = (\tfrac{1}{2})^n L(T) \tag{2}$$

This sequence of triangles approaches some point z_0 which lies inside or on the triangle T. By hypothesis, f is analytic at z_0 and so

$$\lim_{z \to z_0} \left\{\frac{f(z)-f(z_0)}{z-z_0}\right\} = f'(z_0).$$

This means that given any $\varepsilon > 0$, we can find $\delta > 0$ such that if $|z-z_0| < \delta$, then

$$\left|\frac{f(z)-f(z_0)}{z-z_0} - f'(z_0)\right| < \varepsilon. \tag{3}$$

The condition $|z-z_0| < \delta$ means z lies in a disc centre z_0, radius δ. Since the sequence of triangles $T_1, T_2, \ldots, T_n, \ldots$ approaches z_0 and each T_n is half the linear dimensions of its predecessor T_{n-1}, for some N we have T_n lying inside this disc for $n \geqslant N$. Thus for all z on the triangle T_n, $n \geqslant N$, from (3) we have

$$|f(z)-f(z_0)-f'(z_0)(z-z_0)| \leqslant \varepsilon|z-z_0| \leqslant \varepsilon L(T_n). \tag{4}$$

By the Fundamental Theorem of Contour Integration, since T_n is a closed contour, $\int_{T_n} dz = \int_{T_n} \frac{d}{dz}(z)\,dz = 0$ and $\int_{T_n} z\,dz$

$= \int_{T_n} \frac{d}{dz}\left(\frac{1}{2}z^2\right) dz = 0$. Since z_0 is fixed, we see that

$$\int_{T_n} f(z)\,dz = \int_{T_n} \{f(z)-f(z_0)-f'(z_0)\,(z-z_0)\}\,dz.$$

From (4), we have

$$\left|\int_{T_n} f(z)\,dz\right| \leqslant \varepsilon L(T_n).L(T_n)$$

$$= \varepsilon(\tfrac{1}{4})^n L(T)^2 \qquad \text{from (2)}$$

Comparing this with (1), we find

$$(\tfrac{1}{4})^n h \leqslant \varepsilon(\tfrac{1}{4})^n L(T)^2$$

i.e.

$$h \leqslant \varepsilon L(T)^2.$$

But $h \geqslant 0$ and ε may be arbitrarily small. This implies that $h = 0$.

Solutions to Exercises

Chapter One

1. (i) $x^2-y^2+2x+i(2xy+2y)$ (ii) $x(x^2+y^2)^{-1}-iy(x^2+y^2)^{-1}$
   (iii) $\sin x \cosh y+i \cos x \sinh y$
   (iv) $(xe^x \cos y-x+ye^x \sin y)\,(e^{2x}-2e^x \cos y+1)^{-1}$
   $+i(ye^x \cos y-y-xe^x \sin y)\,(e^{2x}-2e^x \cos y+1)^{-1}$
   (v) $\frac{1}{2}\log(x^2+y^2)+i\tan^{-1}(y/x)$ where we choose $0\leqslant\tan^{-1}(y/x)<\pi$ for $y\geqslant 0$ and $-\pi<\tan^{-1}(y/x)\leqslant 0$ for $y\leqslant 0$.
   (vi) $x^2+y^2+i.0$ (vii) $\tan^{-1}(y/x)$ with the conventions of 1.(v) above.

2. (i) $2z+2$ (ii) $-z^{-2}$ (iii) $\cos z$ (iv) $(e^z-1-ze^z)\,(e^z-1)^{-2}$
   (v) z^{-1}.

3. (a) $\dfrac{\partial u}{\partial x}=2x,\ \dfrac{\partial u}{\partial y}=2y,\ \dfrac{\partial v}{\partial x}=0=\dfrac{\partial v}{\partial y}.$

   At $z=0,\ \dfrac{d}{dz}(|z|^2)=\lim\limits_{z\to 0}\dfrac{|z|^2-|0|^2}{z}=\lim\limits_{z\to 0}\dfrac{z\bar{z}}{z}=0.$

   (b) $\dfrac{\partial u}{\partial x}=\dfrac{-y}{x^2+y^2}\quad\dfrac{\partial u}{\partial y}=\dfrac{x}{x^2+y^2},\ \dfrac{\partial v}{\partial x}=0=\dfrac{\partial v}{\partial y}.$

4. (i) domain (ii) no (not open) (iii) no (not connected) (iv) yes (v) yes.

5. $f=u+iv$ where $v=0$. $\dfrac{\partial v}{\partial x}=\dfrac{\partial v}{\partial y}=0$. By the Cauchy-Riemann equations $\dfrac{\partial u}{\partial x}=\dfrac{\partial u}{\partial y}=0$. Thus $f'=0$ throughout the domain of definition which implies that f is constant.

6. (i) $\lambda_1=ik,\ \lambda_2=-ik$ (ii) Ae^z+Be^{2z} (iii) $Ae^z+Be^{2iz}+Ce^{-2iz}$.

7. $(1-z)^{-4}=1+4z+\dfrac{4.5}{1.2}z^2+\ \ldots\ +\dfrac{(n+1)\,(n+2)\,(n+3)}{6}z^n+\ \ldots$

Chapter Two

(In questions marked * the integration may be performed using the Fundamental Theorem of Contour Integration).

1. $-\frac{1}{2}+\frac{1}{3}i$ 2*. -1 3*. $e^{i}-e$ 4*. $\operatorname{Log} i-\operatorname{Log} 1 = i\frac{\pi}{2}$ 5*. $-\frac{1}{3}i-\frac{1}{3}$

6. $\frac{2}{3}i$ 7*. $\frac{1}{2}i-\frac{1}{4}i\sinh 2-\frac{1}{2}+\frac{1}{4}\sin 2$

8*. $\frac{i^{\alpha+1}}{\alpha+1}-\frac{1^{\alpha+1}}{\alpha+1}=\frac{e^{i(\pi/2)(\alpha+1)}-1}{\alpha+1}$ (since $i^{\alpha+1} = e^{(\alpha+1)\operatorname{Log} i} = e^{i(\pi/2)(\alpha+1)}$)

9*. 0 10. 0 11. $2\pi i$.

12. (i) $\sum \frac{c_n z^{n+1}}{n}$ converges absolutely by comparison with $\sum |c_n z^n|$ since $\left|\frac{c_n z^{n+1}}{n}\right| \Big/ |c_n z^n| = \frac{|z|}{n} \to 0$ as $n\to\infty$.

(ii) $\int_\gamma f(z)\,dz = F(z_0)-F(0) = \sum \frac{c^n z_0^{n+1}}{n}$.

Chapter Three

1. $z+2z^2+\ldots+nz^n+\ldots |z|<1$ 2. $z^3+z^4+\ldots+\frac{z^{n+3}}{n!}+\ldots$ for all z.

3. $1+3z+3z^2+z^3$ for all z 4. $1-z+\ldots+(-1)^n\frac{z^n}{n}+\ldots |z|<1$.

5. $1+\alpha z+\ldots+\frac{\alpha(\alpha-1)\ldots(\alpha-n+1)z^n}{nh!}+\ldots |z|<1$ (unless α is a positive integer in which case the series terminates and is valid for all z).

6. $1-z^2+\ldots+(-1)^n z^{2n}+\ldots |z|<1$.

7. If $|z-z_0| = r$ and $r \geqslant |z_0|$, then $|z| \leqslant |z_0| + |z-z_0| \leqslant 2r$ and so $|f(z)| \leqslant 2^n r^n M$. Hence $|f^{(n+1)}(z_0)| \leqslant \dfrac{2^n r^n M(n+1)!}{r^{n+1}}$. Let $r \to \infty$, then $f^{(n+1)}(z_0) = 0$.

8. $\dfrac{z}{(1-z)^2}$ $(z \neq 1)$

9. $\dfrac{z(1+z)}{(1-z)^3}$ $(z \neq 1)$

10. $(1+z^2)^{-1}$ $(z \neq \pm i)$.

11. $f(z) = \dfrac{z}{1-z} + \dfrac{1}{z}$ $(z \neq 0, 1)$.

12. $f(z) = \sum_{n=0}^{\infty} (-1)^n z^{2n} = (1+z^2)^{-1}$ wherever f is defined. Hence $f(0) = 1$, $f(1+i) = (1-2i)/5$, $f(2{,}000) = 1/4{,}000{,}001$ (if they are defined) but $f(i)$ cannot exist if f is analytic.

MEDICAL COMPUTING GROUP
LIBRA Y

INDEX

45p

FUNCTIONS OF A COMPLEX VARIABLE

I

MEDICAL COMPUTING GROUP
LIBRA Y

LIBRARY OF MATHEMATICS

edited by

WALTER LEDERMANN

D.Sc., Ph.D., F.R.S.Ed., Professor of Mathematics, University of Sussex

Complex Numbers	W. Ledermann
Differential Calculus	P. J. Hilton
Differential Geometry	K. L. Wardle
Electrical and Mechanical Oscillations	D. S. Jones
Elementary Differential Equations and Operators	G. H. H. Reuter
Fourier and Laplace Transforms	P. D. Robinson
Fourier Series	I. N. Sneddon
Functions of A Complex Variable 2 vols	D. O. Tall
Integral Calculus	W. Ledermann
Introduction to Abstract Algebra	C. R. J. Clapham
Linear Equations	P. M. Cohn
Multiple Integrals	W. Ledermann
Numerical Approximation	B. R. Morton
Partial Derivatives	P. J. Hilton
Principles of Dynamics	M. B. Glauert
Probability Theory	A. M. Arthurs
Sequences and Series	J. A. Green
Sets and Groups	J. A. Green
Solid Geometry	P. M. Cohn
Solutions of Laplace's Equation	D. R. Bland
Vibrating Strings	D. R. Bland
Vibrating Systems	R. F. Chisnell